Antique Pocket Mirrors

Pictorial and Advertising Miniatures

Schiffer Publishing Ltd ®
4880 Lower Valley Road, Atglen, PA 19310 USA

Cynthia Maris Dantzic

Dedication

For Jerry, who shared our early days of discovery with delight and enthusiasm.

Fig. 88

Library of Congress Cataloging-in-Publication Data

Dantzic, Cynthia Maris.
Antique pocket mirrors: pictorial and advertising miniatures/by Cynthia Maris Dantzic.
p. cm.
ISBN 0-7643-1646-X (pbk.)
1. Mirrors--Collectors and collecting--Catalogs. 2. Advertising specialties--Collectors and collecting--United States--Catalogs. 3. Miniature objects--Collectors and collecting--Catalogs. I. Title
NK5440.M553 D36 2002
748.8--dc21
2002009767

Designed by "Sue"
Type set in ShelleyAllegro BT/Souvenir Lt BT

ISBN: 0-7643-1646-X
Printed in China
1 2 3 4

Please note that some mirrors in this book are shown larger than actual size.

Published by Schiffer Publishing Ltd.
4880 Lower Valley Road
Atglen, PA 19310
Phone: (610) 593-1777; Fax: (610) 593-2002
E-mail: Schifferbk@aol.com
Please visit our web site catalog at
www.schifferbooks.com
We are always looking for people to write books on new and related subjects. If you have an idea for a book, please contact us at the above address.

This book may be purchased from the publisher.
Include $3.95 for shipping.
Please try your bookstore first.
You may write for a free catalog.

In Europe, Schiffer books are distributed by
Bushwood Books
6 Marksbury Avenue
Kew Gardens
Surrey TW9 4JF England
Phone: 44 (0) 20 8392 8585
Fax: 44 (0) 20 8392 9876
E-mail: Bushwd@aol.com
Free postage in the UK. Europe: air mail at cost.

Contents

Fig. 63

Acknowledgments

This book could not have been written without the generous assistance of many people, some of whom I have not met in person but rather communicated with through the mail, telephone, or various computer-related technologies.

My sincere expression of appreciation is extended particularly to those who have permitted me to borrow personal treasures—their parents' wedding portrait, a grandmother's youthful image, their own baby picture or confirmation photo on a pocket mirror. Thank you Florence Baravarian, Sheila Brummel, Marilyn Cuff, Connie Crespo, Juliette Galant, Shifra Gassner, Helene and Frank Manzo, Goldie Mossman, Robert Rindler, and John Todaro, for sharing your own family history.

For making available an extensive collection of dexterity toy or game mirrors, which require skill and patience in manipulation to convince tiny beads to fit into tiny depressions, or other difficult maneuvers of hand and eye coordination, I am most indebted to Christina and Irwin Rubin.

The extra efforts and kindness of Burt Purmell, Roger Steckler, Ted Hake, Mark and Lynda Suozzi, and Dr. Robert Schuman in providing images of exceptional mirrors from their own collections have richly enhanced my more modest assortment of treasures. Thank you also, Burt, for reading this manuscript with your professional eye!

I am particularly appreciative of the assistance of Paul J. Bosco, a dealer in tokens and coins, for sharing his listing of fellow entrepreneurs and collectors, that enabled me to approach many of the major participants in the pocket mirror world. Through Mr. Bosco I have enjoyed correspondence and contributions from such experts in this field as Bill Aquilino, Larry Elman, Rich Hartzog, Chuck Kirtley, Bob Moffat, Rex Stark, Ted Wahlfeld and Alan Weinberg. My thanks to you all. Carl Burnstein, another coin and token dealer in Manhattan, was also helpful.

A difficult task, the assigning of a range of actual value in the current market, was carefully considered and estimated by several collectors and dealers to whom I am greatly indebted. These include Larry Elman, Ted Hake, Chuck Kirtley, Dr. Robert Schuman, and Roger Steckler, among others. For this, I consulted several auction price lists plus annual editions of *Kovels' Antiques & Collectibles Price List*, as well.

Several copies of illustrated articles by Burt Purmell were sent to me by admiring collectors. They are included in the Bibliography at the end of this book. It is well worth the effort to find these gems of writing and imagery through research in antiquarian book shops or even through the Internet, as they are now out of print.

A special word of thanks is extended to the dozen or so members of my weekly "Sterling Drawing Group," which meets at the home of Jo Bidner, for assisting with identification of period fabrics, styles, and other clues needed to pinpoint the dating of many personal pictorial works.

Thank you, Tim Keating, for your patient, professional photographing of much of my collection in groupings on 6" x 9" sheets of Color Aid paper.

I extend particular thanks to my friend and typist, Jeannie Butler, who has undertaken similar tasks for me in the past. Her meticulous, precisionist approach to a job in which every date and comma must be perfection requires patience and a sweet disposition, qualities which she possesses in abundance.

And certainly a word of appreciation goes to Donna Baker, my editor at Schiffer Publishing, Ltd. It is only through her vision and confidence that this work has been brought to its realization.

In many ways, small and large, I am indebted to a long list of others—friends, family, and professional associates, such as Betty Spitz and Alan Kahn—who helped to facilitate the creation of this book. Thank you, most of all, Grayson, Nancy, Irma and Elsy. You lead the list.

Fig. 115

Introduction

Perhaps it is my curiously acquisitive nature and inability to leave behind any well-designed, cleverly worded, or historically interesting specimen that has led to my broadly eclectic collection of pocket mirrors and, inevitably, to this book.

Intrigued by their small size, intricate design, and striking color, as well as their pocket change price tag, my husband Jerry and I first began to notice them in countertop showcases at flea markets and antique shops in Maine some thirty-five years ago. I soon became addicted to the search for these miniature exemplars of American graphic design.

The first in this collection (Fig. 1) was acquired by chance, when the owner of a small antique shop asked, "May I help you, madam?" Hesitant to reply, "I'm looking for a nice geometric Amish quilt," lest she actually have one, putting my entire summer's budget at risk, I noticed a small tan oval shining on the counter and instead said, "Yes, how much is this little . . . um . . . object?" With the payment of a dime, I was on my way!

As you will note, the one-that-started-it-all is a faded memento of the Hotel Royal Poinciana in Palm Beach, Florida, its greenish-tan, tannish-green hues melted with time. I'm showing it here against two background colors to demonstrate how presentation can affect the apparent color and the visual impact of an individual mirror.

Soon I was paying fifty cents, then a dollar apiece—and within a few years, as much as three dollars each, feeling vaguely reckless. One time, Jerry came home with three mirrors (the Starrett Tools, the 1904 Coca Cola, and the Beeman's Pepsin Gum) for which he'd paid a total of ten dollars. I told him to watch it—things were getting out of hand! (Just check the value of these today!)

Over time, I became aware of the great variety, not only of subject matter, but also of materials used in the manufacture of pocket mirrors, and I began to look for specific exemplars of each variant.

Here are ten favorites:

- products in their containers
- products or services featuring women's faces or figures
- smiling faces that turn to frowns when inverted
- mirrors whose oval or round shape becomes the contour of the product itself
- mirrors that are good for ten cents or more, in trade or when shown (these are known as "good fors")
- brass or silver-plate inkless, embossed-top circles (early examples)
- printed paper discs, or paper under glass, with tin loop edges (also pre-turn of the century)
- brides and grooms, and other personal portraits
- political or patriotic subjects
- boats, trains, cars or other means of transportation

Invariably, other collectors, often dealers encountered at flea markets or shops, will register approval or disinterest when certain subjects or types of mirrors are mentioned. Early paper disc mirrors are of little interest to those who seek the color and strong design found in celluloid examples. However, they may be of great historic concern to the devotee of pre-turn of the century typefaces or of such obsolete products as horse blankets, roller tray trunks and grate bars for coal, the services of a "hatter and shirter," or H. Fischer's Pool Room in Niagara Falls!

The vast field of personal pictorial imagery is rich in elaborately attired bride and groom portraits, nicely displaying the appropriate dress of their decade and the relative position of the two figures within their oval, round, or rectangular field. Collectors of advertising mirrors often pass these by. However, if the name and location of the portrait studio is shown, then they, too, are advertisements! On the other hand, students of the human experience, of the social artifact, of early twentieth century portrait photography often find these the most appealing.

I am an equal opportunity collector; after all, duplicates or pieces that lose their appeal can always be traded away, later . . .

Aims & Uses of the Book

At this point, you can well appreciate my interest in preparing the book you are about to enjoy. One of my major concerns is "to open eyes"[1] to a little-known, but growing area of American pictorial and advertising design history.

Art historians, artists, designers, and students will find much that speaks to them in these pages. For example, decades before Andy Warhol immortalized the Campbell's Soup can and other icons of everyday life as part of the Pop Art movement, the humble precursors shown in Fig. 121 surely set the pace!

And, aren't these mirrors beautiful! They were made with skill, meticulous craftsmanship, superb application of color, design, and lettering or typography, and often, unwittingly, with an apt touch of humor through the excessive zeal of American entrepreneurial salesmanship. Exaggerated claims for a product, such as "cures a cough in 1 day, and the croup in 1 minute" or "for the sick and weakly especially Nursing Mothers," are matched by such appropriate "pre-Hallmark" verses as:

> If you the only rival of
> the B-ROSE SWEETS would see
> just turn this little souvenir
> and the answer there will be.

Perhaps you are already a pocket mirror aficionado, with several dozen, even several hundred, examples in your collection. The more familiar you are with this subject, the more you will understand that no single book can include all the designs that have been produced. Unlike stamps, for instance, whose collectors may hope to acquire, say, every American stamp published after a certain year, the vast number of small volume, even individual, mirrors produced throughout the United States (or printed in Germany for distribution here), precludes that possibility. And through the years many have been discarded, destroyed, or worn out through use.

I am fortunate in having been granted permission to include images of some of the most highly desirable mirrors, the early Coca Colas in Roger Steckler's collection, for example, and the many rare and excellent pictorial advertising pieces in the collections of Burt Purmell and Robert Schuman. But, even collections of several thousand specimens (yes, I have discovered a number of these national treasures) cannot be all-inclusive or complete. And, if they were, you could count on their owners to continually seek ever more perfect exemplars of each one—the search is never-ending!

What this book will provide, short of that elusive, comprehensive impossibility, is a rich overview of the entire field. It presents a brief background of pocket mirrors and the specific manufacturers who produced them, plus as complete a categorical listing of materials, types, and subjects as possible. Also provided are suggestions to assist in locating mirrors, whether at the beginning of your search or at a more sophisticated level, and specific advice regarding such considerations as:

- authenticity: avoiding fakes
- condition: what to look for
- conservation: how to preserve
- repairing: how to remove flaws
- identifying: date, location, artist
- displaying: to best advantage
- value: how to estimate
- research: how to find the facts

A glossary is included to help with the understanding of specialized terms used by dealers and collectors. "I see there's some foxing on the curl here," just may convince a dealer that you're not a novice and can help when you are discussing the price.

The Bibliography will direct you to several compendiums covering specific collecting areas as well as articles that have been published. Though out of print, they may still be located and are worth the effort involved. This is the first book devoted to the subject in general.

The Bibliography also provides specific auction lists to which you may subscribe. These are good sources for acquiring pocket mirrors and include the names and addresses of a number of dealers you may wish to contact.

There are many places to look for mirrors and you never know what will appear in the very next showcase or on the very next table. Happy hunting . . . As Sherlock Holmes would say, "The game's afoot!"

[1]Josef Albers's oft-quoted reply, when asked his goal as a teacher of color and design.

Fig. 117

Historic Background

The pocket mirror in America served several definite purposes around the turn of the last century, some commercial, some social, all aesthetic.

> Decades before television made "commercials" a component of everyday American life, 20th-century advertisers attracted attention to their wares by distributing utilitarian gadgets embellished with clever slogans and eye-catching illustrations. Popular giveaways included tin trays, hand-held fans, and calendars adorned with images of nubile young women and cherubic tots. Equally handsome, though less common, were celluloid-backed pocket mirrors. In vogue from about 1900 to 1930, these close cousins of the pin-back button relied on the magic of chromolithography to portray a positive image of a product or service. Oval, round, or rectangular, the mirrors ranged from about an inch and a half to five inches wide and tucked neatly into a purse or vest pocket. Most important, as far as advertisers were concerned, was the notion that each time the owner assessed his or her grooming in the looking glass, he or she would receive a subliminal reminder to buy the brand of cooking oil that yields the "best pastries" or the trolley-car fender that "saves human lives." Today, collectors prize these lithographed treasures for their nostalgic appeal, graphic content, and the color and complexity of their illustrations.[2] (Fig. 2)

As early as 1868, a patent was issued for an inkless, embossed brass-topped shell mirror. Often backed by a paper disc, another brass design, or even a coin replica, this brass-edged shell was the first commercial object to include a small pocket-size mirror. Its wide rim was often pierced with holes to hold a number of straight pins, thus providing a third function: it served as advertisement, pocket mirror, and pin-holder as well.

Less costly to produce, the silver- or nickel-plate embossed shell mirror soon followed. In later years, aluminum was sometimes used (see the Maxwell House Coffee in Fig. 5, made in Germany for the American market).

A natural successor was the extremely inexpensive paper-disc-faced mirror, attached to a pasteboard circle or oval and held in place by a thin tin edging loop. Extremely fragile and susceptible to damage, the paper disc was soon covered by a thin sheet of glass. Even these proved delicate and permitted limited freedom of color and design.

The introduction of clear celluloid, around the turn of the century, brought a radical change. At first, a simple wrapping of this transparent material over the folded edge of a rectangular printed image seemed quite an innovation. Textured papers and other fabric-like surfaces, sometimes a metallic gold-like foil, appeared. Then, everything changed, when that thin sheet of celluloid was placed over an intricately printed, colored lithographic image (chromolithography), held in place (often over a cardboard form) and backed by a small mirror, with everything clamped together by a metal ring or framing loop called a "collet." This allowed the celluloid to encase the mirror completely, providing a rim or curl upon which the manufacturer's name could be printed (often in excruciatingly tiny type, requiring a magnifying glass to read). It is always a delight to find the occasional offer of a duplicate mirror "for 6¢ in stamps." If I could only obtain one of those today!

Soon, nearly all pocket mirrors were made of celluloid—actually pyroxylin (cellulose nitrate) and camphor. The two names associated with the development of this versatile, flexible, and inexpensive material are John Wesley Hyatt and Anselme Payeu. Hyatt was the American inventor who, in his search for a substitute for ivory, experimented with cellulose. In 1870, he obtained a patent for "celluloid," the first successful synthetic material and the precursor of all the plastics we use so freely today. His work followed that of Payeu, the French chemist who was first to identify cellulose itself in 1839.

Of course, celluloid has been used to make a wide variety of products in addition to pocket mirrors. Pin-backed political and other buttons, measuring tape containers, whetstones for blade sharpening, velvet or plush-surfaced record dusters, calendars, notepads, bookmarks, shoehorns, dance card holders, stamp cases, watch fobs, and badges were among the most popular.

Often, the same stock images were made available for use on several products. You will find the same Victorian ladies pictured on more than one object, for example, or advertising more than one business or product on the same mirror blank.

Hal V. Dunn's *Catalog of Western Good for Trade Mirrors* (see Bibliography), pictures 130 different stock images or "types" (mainly women and children). Shown here are over a dozen of them, not all on "good for" mirrors (which, according to Dunn, were all issued before 1930, mostly around 1908).

The first major manufacturer of celluloid products was the Whitehead and Hoag Company of Newark, New Jersey, founded in 1892. Until this firm was sold to its chief competitor—Bastian Brothers of Rochester, New York—in 1959, it led the field, having patented celluloid pin-back buttons in 1896 and 1898. Production of buttons was soon followed by that of pocket mirrors and many of the other small advertising items previously named. A detailed biography of Benjamin S. Whitehead and Chester R. Hoag, including the checkered labor history of their famous company, may be found in Hake and King's *Price Guide to Collectible Pin-Back Buttons 1896-1986* (see Bibliography).

The Bastian Brothers, Theron E. and Frederick J., were jewelers in Rochester who started their company in 1895. They remained Whitehead & Hoag's chief competitor until they purchased that business, as noted above, in 1959. The merger involved the loss of many historic Whitehead & Hoag records and documents, though Bastian continued to use the Whitehead & Hoag name on some of its products until 1965.

Most of the mirrors shown in this book were made by one of the "Big 5" companies, whose name will invariably appear on the rim. In addition to Whitehead & Hoag and Bastian Bros., these are the Parisian Novelty Co., Chicago, Illinois; Cruver Mfg. Co., Chicago, Illinois; and The American Art Works, Coshocton, Ohio. Of the many other smaller companies, too numerous to list, you may occasionally find an example made by one of these firms: St. Louis Button Co., St. Louis, Missouri; The Meek Co., Coshocton, Ohio; and Lou Fox, Chicago, Illinois.

Advertising pocket mirrors were generally given out by product manufacturers or by individual shopkeepers (along with the product at the time of purchase), as premiums or simply as keepsakes to remind the customer to return. As noted earlier, duplicates might be available by mail for as little as six cents in stamps. "Good for" mirrors could be exchanged (or sometimes shown) for a discount, a cigar, or a drink.

To produce a smaller reflecting image size, but over a considerably larger field, a convex reducing mirror was sometimes used as a variant to the standard mirror glass. In this book, pocket mirrors containing reducing mirrors can be seen in Figs. 26, 31, 87, 145, and 154.

[2]Written by Burt Purmell, for his article in the June 1992 *Country Living*, this is a beautiful introduction to the subject. (Thank you Burt, for permission to include it here!)

Fig. 138

Collecting Pocket Mirrors

If you are just beginning a collection of pocket mirrors, this section will aid you in your search as well as in the preservation and display of your treasures. Long-time collectors may wish to skim this material, if only to verify the information or, perhaps, to send along favorite tips and suggestions for inclusion in a future edition!

What to Collect

Pocket mirrors can be grouped into categories by type of material, subject, location, or specific design elements. You may decide to limit your collection by focusing on one or more of these categories.

By Material

Embossed metal shell: brass, nickel or silver-plate, aluminum
Paper-topped
Glass-covered paper
Textured paper under celluloid
Fabric-like materials
Celluloid (over chromolithograph or photograph)
Other materials (tin, wood, leather, and rubber, rarely seen)
With handles: rigid or folding
Mirrors as container tops (compacts)

By Subject

This is, of course, a very subjective area, since the same mirror may be classified differently according to a collector's personal criteria. For instance, the famous Victor "His Master's Voice" (Figs. 117 and 123) might be grouped under "Music," "Non-electric machines," "Dogs" or "Classic products." For display purposes, you may need more than one!

There are many ways to group your collection by subject. Ted Hake and Russ King, in assorting pinback buttons, use these thirteen categories:

Animals, birds, creatures
Celebrations, expositions, World's Fairs
Entertainment
Famous people
Farm equipment
Food, beverage, tobacco
Geographic areas
Military
Organizations, associations
Products, services
Social causes
Sports
Transportation

Although this listing seems inclusive, it omits several categories of possible interest to a collector of pocket mirrors, rather than pinback buttons. Here is a list I prefer, which can be modified to suit individual predilections:

Advertising

Illustrated:
Product image the main element
Stock image or one of a kind
Image other than the product (e.g., women, shown with or without product)
Lettering only, no image
With actual coin inclusion beneath the celluloid (see Fig. 169).

Products Specifically For Use By:
Women
Children
Men

Product Categories:
Food (in tins, boxes), drink
Clothing
Medicinal, pharmaceutical
Publications
Recreational

Birthstones Encircling Field:
May be further subdivided as pictorial, jewelry, type only . . .

Related Circumferential:
Clock faces, calendars, measuring tapes . . .

Animals:
Farm, domestic (can cross into other categories), wild

Reflection-related:
("The lady on the other side is requested . . .")

Shaped Image:
(Entire mirror is the product, i.e., barrel, egg, football . . .)

Inverted Image:
(Smile becomes a frown, Naughty Nellie, q.v.)

Theatrical:
Actors, theaters, plays, films

Transportation
Trains, ships, boats, canoes, cars . . .

Political:
Candidates for office, elected officials
Civic events
Patriotic themes, use of flag, red/white/blue
National monuments, parks
Military

Geographic:
Specific locations by state, city, town . . .

Buildings:
Factories, churches, historic sites, homes . . .

Holidays:
Christmas-related, Fourth of July . . .

Good fors:
Good for specific amount in trade

Recreation:
Dancing, skating, boating

Music:
Instruments, performers, related products

Pictorial

Wedding Portraits:
Bride alone, bride and groom, groom alone . . .

Family Ceremonies:
Non-wedding, confirmations, bar mitzvahs, children alone or with parents, graduations

Personal Documents:
Summer vacation, at beach, mountains, in garden

Personal Portraits:
Men, women, children, pets . . .

Manipulated
Dexterity games or toys, with beads, dice, other moveable elements to be controlled
Flip-flops (hinged illustrations that reveal less-clothed figures when "flipped")
Naughty Nellies (reveal risqué image when inverted)

Paperweights
Larger, heavier versions of standard pocket mirrors for desktop use

Other Than Mirrors

Whetstones:
Instead of mirrors on reverse, these are fitted with honing or sharpening stones

Record-dusters:
Velvet or plush-covered padding, usually paperweight size

Tape measures:
Smaller than pocket mirrors, showing identical image on both sides, usually portraits of Victorian women

Your own additions or modifications are invited. Have I omitted your favorite?

Making Your Choice

In addition to satisfying your personal criteria for subject matter, there are other considerations that will help you determine which mirrors to acquire and which to leave on the table.

Authenticity

How can you detect an outright fake, reproduction, or fantasy mirror? (Consult the glossary if you are not sure of the differences.) Your best guarantee of authenticity is to patronize reputable dealers. Reputations depend on dealers' reliability. Ask questions of these professionals; it is the best way to increase your knowledge of the subject. The following are a few suggestions and guidelines.

If you find a Coca Cola mirror, watch out. You are unlikely to see an authentic one of these rare and highly collectible gems! However, in the 1970s, the Coca Cola Company issued dated and acknowledged reproductions, some as key chain attachments; they are certainly not valuable, but not fake either.

To detect a reproduction or fake, first check the reverse and edge to see if the celluloid fits the collet ring smoothly. If folds and wrinkling are evident, they are sure signs of a poorly-made imitation.

Another clue—check the covering. Modern acetate may have been used instead of celluloid on mirrors produced since the 1950s and on reproductions. (This is not possible to detect on Internet auctions such as eBay.)

Use a magnifying glass to spot the telltale dots of a BenDay screen printing job. The early stone lithographic process produced no such "texture," one of the reasons its colors are so sharp and clear. Actual photographs also reveal no screening.

Absence of certain details you know to exist on a bona fide mirror can be another indication of a reproduction. See the two Garland Stove mirrors (Fig. 5); they are not quite identical. Which do you suspect of "inauthenticity" (And, why?).

In addition to the myriad Coke copies, you may find other popular brand imitations, such as Hires Root Beer and Pepsi Cola. Check the rim for an identifying maker; check the curl for smoothness; get out your magnifier . . . *caveat emptor*! Buyer, beware!

Condition

You've identified a mirror you would like to buy, you are certain it is authentic, but:

- What about that brown stain at the edge?
- The surface is dull and caked with dirt spots . . .
- The surface is cracked, split or peeling . . .
- The mirror itself is cracked or missing!

The brown stain, called "foxing," is rust; and, if it is underneath the celluloid, it cannot be removed. Careful rubbing with a damp cloth, a little soap, and possibly a touch of liquid wax may remove light foxing on the surface. A mild toothpaste can also work. Be sure to prevent any moisture from entering at the collet's edge, however. Dirt on the surface can be removed the same way. Often a layer of grayness will vanish, revealing brighter, clearer color, even on a mirror not thought to be dirty. Toothpaste will often add an attractive sheen. The secret is to polish rather than bleach a stain away. A drop of acetone is sometimes used. Light scratches can be filled with beeswax or a little soap, rubbed in well. For a mirror that has gold-inked lettering applied to its surface, be careful not to clean too heavily—the gold can be easily removed. Always dry a cleaned mirror thoroughly.

Dents and bumps can, in theory, be removed by an expert, who can disassemble the mirror and smooth the printed image. However, unless you have a rare treasure, this hardly seems worth pursuing.

Cracks in the celluloid, resulting from heat, moisture, age, or poor production procedures cannot be corrected. You are more likely to find such damage on personal, one-of-a-kind pieces, produced perhaps by a local wedding photographer, than on those made by a major company. If the mirror has been exposed to sunlight, radiator heat, or dampness, splits and cracks may occur regardless of its original condition.

Although a cracked, broken, or missing mirror may seem a major impediment, this defect can readily be addressed by a professional. The mirror, unless a convex reducing type, can be replaced invisibly. However, advanced collectors often reject an imperfect mirror.

What determines the degree of imperfection acceptable to a collector? Flaws that can be removed without a trace are not a problem. Certain distinctions should be considered. The rarer the item, the less critical is its condition. Damage to the "field," the area around the main image, is less important than any flaws within that central "portrait" area. Remember to check the edge (curl) as well as the front and back for damage. Faded color cannot be corrected, though buffing with a drop of liquid wax may help.

Value and Desirability

While value is a subjective determination to a point, there is a generally accepted price range that most dealers and auctioneers agree upon for mirrors assumed to be in good condition with no apparent flaws. In order to determine that range for items shown in this book, I consulted with a number of experts and was pleased to find a reasonably consistent evaluation.

A few considerations regarding value were advised by these consultants and I am pleased to pass them on to you.

- Certain terms or words, particularly on an early mirror, can add appreciably to its value—"saloon," for example, instead of "bar" or "tavern." "T" (for Territory) after western state names, such as Arizona or California, Alaska and Hawaii, are very rare and valuable. Western state names are more desirable than those of Eastern states and even Midwestern state names are preferable to those in the Northeast. Basically, the larger the population of a named city, the less rare and less valuable are its mirrors. "Wyoming is better than Colorado," I was advised. "Maine is better than Massachusetts."
- everything you see counts; a seven-digit phone number (pre area codes!) is a sign of the 1940s.
- A stock die is less valuable than individually produced artwork.
- A known illustrator adds to the value of mirrors with Victorian women's portraits. (See Fig. 98, where six illustrations by the famed Angelo Asti, known as "Asti Girls," are shown.) Also check the mirror descriptions for Fig. 96. These are from 1903 to about 1918.
- Certain subjects are considered particularly popular among collectors. In addition to the geographic locations already noted, these include: Product and occupation-related mirrors with alcoholic and smoking subjects; soft drinks such as Coca Cola, Pepsi, Hires Root Beer, 7-Up and White Rock; medical and pharmaceutical subjects, such as patent medicines; doctors and dentists, their services and products. Also included in this category are funeral parlors; pictures of electro-magnets; specific groups of people, e.g., African Americans, any children or babies; animals (domestic, farm, or wild); risqué or erotic subjects, and *chats noir* (black cats). "Some people look for Havana cigars; watches, rather than shoes; ethnic surnames; ships, whether ocean liners or riverboats . . .," my consultant waxed enthusiastically, "a shell mirror with a circus would be a great find!" "Basically, the more on the mirror, the more it cost to make, and thus the greater its value; the more art, more color, more wording . . . more information," he added (see Fig. 2).

To most collectors, value is affected by condition, but desirability may be a different matter. If you need a particular piece to complete a set or to increase your number of "dogs," "good fors," or "Coca Cola" mirrors, you may be willing to overlook a flaw or two, hoping to find a more perfect specimen later and "trade up."

If you were collecting coins, you could use the traditional categories of condition: mint, "like the day it was made"; near mint, "lacking its full luster or brightness, otherwise perfect"; extra fine; very fine; fine; very good; good; fair; and poor . . . in descending order, to set a dollar value.

With pocket mirrors, many other factors can affect your decision to buy . . . as well as the amount you are willing (if not too eager) to pay!

Identification and Research

Here are a few tips to help with identifying and dating pocket mirrors:

- Period telephone directories can help date and locate companies, products, and sources no longer in business or available. Larger libraries may have a good file of these.
- Bride and groom portraits proclaim their decade through style of gown, headdress, suit coat or tie. Other photo pictorials will also have clothing and hairstyle cues.
- A date may be discovered, through use of a magnifying glass, on the face of the mirror or its edge.
- A patent date may be several years earlier than the actual date of the mirror.
- Research can help; check such reference works as Hake and King's *Price Guide to Collectible Pin-Back Buttons 1896-1986* or Kaduck's *Collecting Advertising Mirrors* to find a mirror similar to one you wish to identify.
- To identify stock images, specific artists, or models, refer to Dunn's *Catalog of Western Good For Trade Mirrors*, which pictures 130 separate illustrations. Perhaps you will find Dunn's "types 12 and 13" (see Fig. 90, which shows the beautiful Evelyn Nesbit, "the girl in the red velvet swing," whose marital and amorous exploits made headlines in New York at the turn of the century.) Soon, you should be able to identify the illustration style of Angelo Asti, whose works appear as Dunn's types 122 and 124. Asti created four of the six "Gorgeous!" women in Fig. 98 and several in Fig. 96. About a dozen of the stock types enumerated by Dunn appear here in Figs. 90, 94, 96, 97, and 98.

Display and Conservation

Displaying and Storing Pocket Mirrors

One of the most attractive features of the pocket mirror is its diminutive size, permitting even a relatively large collection of several hundred pieces to be displayed in very little space. Some of the most popular choices for presentation, display, or storage are:

- Black paper-edged flat coin and pin-back button trays made of heavy board with clear plastic tops, most frequently seen at antique shows and flea markets. They are light in weight, filled with protective cotton batting to hold the mirrors securely in place, and may be stacked easily for transport. Top and bottom are held together by straight pins.
- Wooden, flat drawer cabinets made for display and storage of coins, medals, and tokens. These may be made of oak, cherry or other hardwoods and feature brass fittings, felt drawer liners, and other niceties. Expensive, but elegant; mine is ten drawers high.
- Jewelry shop trays that are stackable, open-top, wooden-framed square boxes, often with inner dividers and felt or velvet surfacing. I found a dozen in a Chinese jewelry shop that was going out of business years ago, and have them stacked six-high on each side of my dresser.
- Spool thread metal cabinets made of thin, strong enameled metal, often decorated with sewing products and symbols, usually four narrow drawers high, with glass-fronted panels. I have placed a Star Twist atop a Star Mercerized on a second dresser, to hold about half my collection. In these cabinets, because of their narrow lane-separating strips, it is necessary to put a flat surface inside each drawer; felt-covered matt-board or illustration board works nicely. There is plenty of space. Both Star cabinets were made for the American Thread Company.

Select the color of background material—such as felt drawer liners—carefully, to complement or bring out the colors in the group of mirrors being displayed (see Fig. 1 and 166). Often the ground color will be seen to affect the colors of the subjects themselves.

In Ellen Liman's *The Collecting Book*, there is a photograph of "400 advertising mirrors, hung with double-faced 1/16 inch foam tape, on the walls of [Burt] Purmell's small foyer." In this New York apartment, these were in close formation, loosely in rows and columns, not quite touching and, . . . spectacular! (Burt no longer uses this display method!)

Protecting Pocket Mirrors

Be sure to keep your collection away from heat (not too near a radiator); sunlight (away from an uncurtained window); and dampness or moisture (not in the cellar or bathroom, and dried thoroughly after cleaning). Wrap them individually when carrying or transporting more than one. Be gentle; dropping can crack or break the glass. (Alas, I've done it!)

Pocket Mirrors – The Color Plates

The One That Started It All

Fig. 1. Hotel Royal Poinciana, Palm Beach, Florida, paid 10¢, c. 1960, $50-75 current value (shown on two backgrounds to bring out coloring).

Fig. 2. Star Egg Carriers and Tray Manufacturing Co., Rochester, New York, "J.B. Carroll, Chicago" on edge. $125-175.

Fig. 4. Thea-Nectar, Pure Chinese Tea, "Pat'd June 16, 1868" on face, Chinese characters translated, top: Camellia, or Thea, Sinensis (tea plant); bottom: Tea House, $125-175.

Fig. 3. *Top to bottom, from left: Column 1,* The New Home Sewing Machine, Boston, Massachusetts (pin-holder rim), $50-75; R. C. Pyle, Merchant Tailor and Dealer in Fine Ready Made Clothing, Easton, Pa., Pat. June 16 (18)68, $50-75; Louis Friendly & Co., Clothing, Hats & Furnishing Goods, Worcester, Massachusetts (pin-holder rim), Pat June 16 (18)68, $50-75; *Column 2*, Ireland & Matthews Mfg. Co., Detroit, Michigan (copper; heavy beveled glass mirror), $40-50; Tansill's Punch 5¢ cigar (pin-holder rim), $75-100; C. Foster Up Town Boot & Shoe Emporium, "898 3rd Av Bet 54 & 55 Sts" (NYC?), $50-75; *Column 3*, N. Espenscheid, Hat Manufacturer, New York, "Patent June 16 (18)68" (pin-holder rim), $50-75; Colgate & Co., Toilet Soaps, New York, "Pat. June 16, 1868," $50-75; M. C. Conner Diamond Merchant and Dealer in Fine Watches, Jewelry and Silverware, Burlington, 10 Washington (pin-holder rim), $50-75.

Fig. 6. Embossed silver-tone clock face, $25-40.

Fig. 5. *Top to bottom, from left: Row 1*, "H & W" waists (embossed figure of child), $35-60; *Row 2*, Garland Stoves and Ranges, dull finish, "The Whitehead & Hoag Co., Newark, N.J." on face, $30-40; LaFrance, Shoes for Women, Trimble Bros., Calais, Maine, $35-40; same as first mirror in row, bright finish (no maker's identification), $30-40; *Row 3*, Albarene Stone, Chicago, New York, Boston, dull finish, $25-40; Maxwell House Coffee (Germany), $30-40; same as first mirror in row, bright finish, $25-40.

Fig. 7. *Top to bottom, from left: Row 1,* Lawrence Fine Tailoring, Portsmouth, New Hampshire, $25-35; *Row 2,* David W. Whittier, Raymond, New Hampshire (rubber stamp), $25-30; Continental Tailoring Co. (Clinton, Indiana), hand signed, "B. H. Morgan, Clinton, Ind." $25-35; *Row 3*, H. Fischer's Pool Room, Niagara Falls South, Canada., Tor. (onto) Litho Co (see Fig. 29), $35-50; *Row 4,* Hurst Co. (1910 written by hand), with image of buggy (see Fig. 114), $35-45; *Row 5,* C. F. Cushing, New Bedford, Massachusetts, Roller Tray, Trunk, $25-35; Lemen Bros., Hagerstown, Maryland, Roller Tray & Trunk, $25-35.

Early Paper, Blue

Fig. 8. *Top to bottom, from left: Row 1,* Maines & Bonnallie, Clothiers, Lewiston, Maine, $25-35; Meigs & Company, Head-to-Foot Clothiers, New Haven (Connecticut), $25-35; *Row 2,* Ozonos only odorless, non-poisonous disinfectant, Boston Chemical Company (Massachusetts), $25-35; *Row 3,* 5/A Horse Blankets, 250 styles, $40-50; D. B. Monroe, Boots, Shoes, Trunks & Bags, Middleboro, Massachusetts, $25-35; *Row 4,* Holland & Havener, Bicycles & Sporting Goods, Worcester (Massachusetts), $35-45; *Row 5,* Electric Lustre Starch, $25-35; Lynch & Proctor, Stylish Clothiers of Springfield (Massachusetts), $25-35; *Row 6,* H. B. Close, Clothier, Moravia, New York, $25-35.

Fig. 9. *Top to bottom, from left: Row 1,* Hector Havana Cigars, $35-50; *Row 2,* J. A. Moore, Clothing, Hats, Caps and Fur Goods, St. Johnsbury, Vermont, $10-25; *Row 3,* Claremont Cash Steam Laundry, Claremont, New Hampshire, $10-25; *Row 4,* Meyer M. Levy's, Clothing, Bangor, Maine, $10-25; W. K. Brooks Coal, Worcester (Massachusetts), $10-25.

Fig. 10. *Top to bottom, from left: Row 1,* Russel W. Parish, Jeweler, Rutland, Vermont (see Fig. 11), $25-30; *Row 2,* Maines & Bonnallie, Clothiers, Lewiston, Maine, $25-30; W. C. Goodwin, Good Shoes, Fitchburg, Massachusetts (not a mirror, but reverse side has dexterity puzzle of a woman's high shoe with black bead buttons; see Fig. 77, top), $25-35; *Row 3,* Ross, Hatter & Shirter, 56 E. Main St. (no city or state), $25-30; *Row 4,* Nathan Daggett, Shoes, Dexter, Maine, $25-30; Gonyea & Aiba, Complete House Furnishers, Rutland, Vermont, $25-30; *Row 5,* Thing's Shoes, Hudson, New York, $25-35; *Row 6,* H. E. Travis, Jeweler, Easton, Pennsylvania, $25-30; Rutland Outfitters, Rutland, Vermont $25-35.

Fig. 11. *From left:* Same as Russel W. Parish mirror in Fig. 10; shown with celluloid Parish mirror, after he had moved to Burlington, Vermont, and added, "Jeweler and Optician" to his title.

Fig. 12. *Top to bottom, from left: Row 1,* The Richmond Stove Company, Norwich, Connecticut, New York, New York, glass over paper, $30-45; *Row 2,* Ballards Obelisk Flour, Louisville (Kentucky), paper, $30-45; Syrup Pix Liquida Compound, $30-45; *Row 3,* W. W. Tupper & Co., New York, New York, Grate Bars for Coal, paper, $30-40; *Row 4,* 5/A Horse Blankets, paper, $30-40; N. Snellenburg & Co., Philadelphia (Pennsylvania), paper, $25-35.

Fig. 13. *Clockwise, from top:* A souvenir from the Universal Theatres Concession Company, Chicago, Illinois, with wire handle (made in Germany), $25-40; Hackley & Moran, furniture, carpets and bicycles, Brattleboro, Vermont (see Fig. 28), $35-50; "Follow the Crowd to the Westchester Clothing Company," Third Avenue and 144th St. (New York?), $25-30; Burlington Steam Laundry, C. R. Huntley, Proprietor, Burlington, Vermont, $25-30; Teen Age Bags, P.I.G. Co., New York (young woman in feathered hat; paper wrap-around), $25-30.

Fig. 14. Cycle Cigarettes, "Manufactured by British-American Tobacco Co. Ltd., United States of America" on face, $100-125.

Fig. 15. *Top to bottom:* Mrs. Fay S. Betger, Hemstitching, 942 East 75th Street (no city, state), on vellum surface paper, in tabbed metal collar with metal strip handle, \$25-35; H. Stepper, Practical Ladies' Tailor, Brockton, Massachusetts, on vellum surface paper, metal collar, \$25-30.

Fig. 16. *Top to bottom, from left: Row 1,* Newman Bros. Co., Pianos and Organs, Chicago (Illinois), Printed in Germany" (see Figs. 83, 97, 123, 124), $25-35; *Row 2,* Pond's Extract, Pond's Extract Company, New York and London, $25-30; The Common Sense Shoe Store, Reading, Pennsylvania, Signature, S. Schweriner, $25-30; *Row 3,* Star Soap, Schultz & Co., Zanesville, Ohio, $25-30; *Row 4,* King's Pure Malt, Pure Food Department, Boston (Massachusetts), $25-30; Colby's Clothing House, Taunton, Massachusetts, $25-30.

Fig. 17. *Top to bottom, from left: Row 1,* Speer's Wines, Passaic, New Jersey, "For the Sick and Weakly especially Nursing Mothers . . ." (Made in Germany), $25-30; *Row 2,* Wentzell & Weber, Clothiers and Furnishers, Bridgeton, New Jersey, $25-30; Sourwine, Kuhn & Kerschner, Clothiers, Allentown, Pennsylvania, $25-30; *Row 3,* Mrs. Dinsmore's Cough and Croup Balsam, L. M. Brock & Company, Lynn, Massachusetts, $50-70; *Row 4,* May Manton, Shoes for Women, Roth Shoe Mfg. Company, Cincinnati, Ohio (Made in Germany), pin-holder along edge, $25-30; *Row 5,* German Heater (no glass covering), pictures heater (see Fig. 153), $40-60; Duluth Imperial, Imperial Mill Company's Patent 196, Duluth, Minnesota, $25-30; *Row 6,* Spicer Stove Company, Providence, Rhode Island, Model Ranges & Heaters, $25-30.

Glass Over Paper, Printed In Reds

Fig. 19. Carson Gloves, "Made by Skilled Labor," $40-50.

Fig. 18. *Top to bottom, from left: Row 1,* Old Crane Whiskey, $25-40; *Row 2,* I. A. Graves & Co., Watertown, New York, White Eagle Flour, $25-35; The Robert A. Wooldridge Co., Baltimore, Maryland, Wooldridge Fertilizers, $25-35; *Row 3,* A. B. Smith & Co.'s 3-20-8 Extra Good Union Hand Made 10¢ Cigar, Boston, Massachusetts, $40-50; *Row 4,* Silas Peirce & Co., Boston (Massachusetts), Peirce's Nine, 5¢ Cigar, $40-50; "The Acton" 5¢ Cigar (no location), $40-50; *Row 5,* McCormick, $60-70.

Fig. 20. *Top to bottom:* Textured fabric-like material (see Figs. 26, 27), Post's Worcester Business Institute (Worcester, Massachusetts), Pat. 11-2-1915, 6-26-1917, Parisian Nov. Co. Chicago, mirror recessed and set away from edge, served as container lid (see Figs. 27, 169), $20-45; Printed on gold foil paper, Becker's Grove, Barn Dance, Mt. Bethel, Plainfield, New Jersey, $15-25; Gold ink on suede-like surface, Patterson Funeral Home, Perryville, Maryland, $25-45.

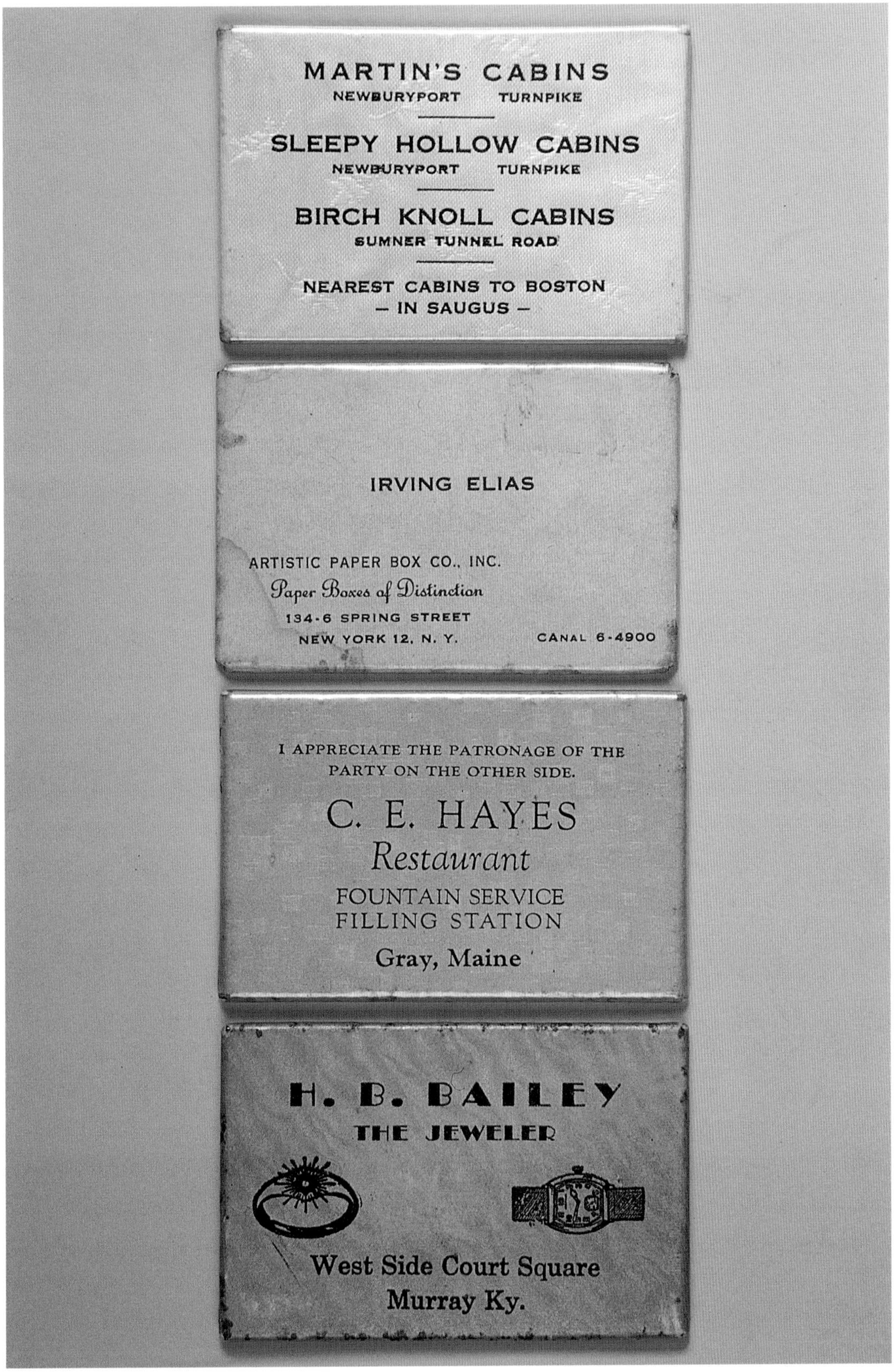

Fig. 21. *Top to bottom:* Martin's Cabins, Newburyport Turnpike, Sleepy Hollow Cabins, Newburyport Turnpike, Birch Knoll Cabins, Summer Tunnel Road (Massachusetts), $15-35; Irving Elias, Artistic Paper Box Co., Inc., New York 12, New York, Advance Mirror & Novelty Co., Inc; C. E. Hayes, Restaurant, Gray, Maine, $15-25; H. B. Bailey, The Jeweler, Murray, Kentucky, $20-35.

Fig. 22. *Top to bottom:* Coney Island (Brooklyn, New York), c. 1940s, $30-50; Tujague's Restaurant, est. 1856, New Orleans, Louisiana, $10-20; Georgia Safety Fire Comm., with 1961, 1962 calendars, $10-20.

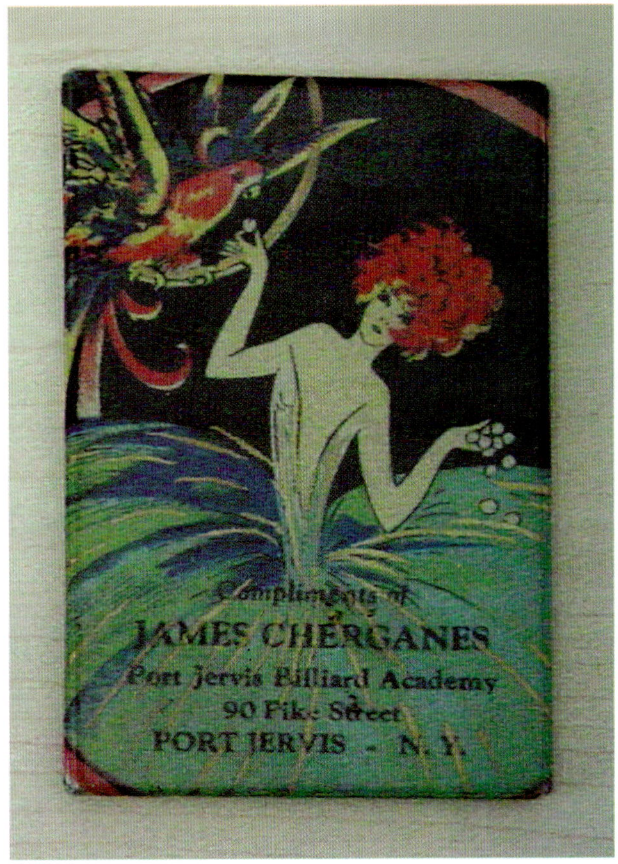

Fig. 23. *Top to bottom, from left: Row 1,* Albert Ritter, Jeweler, Buffalo, New York, $50-70; James Cherganes, Port Jervis Billiard Academy, Port Jervis, New York, $50-70; *Row 2,* Joyson's Dresses Inc., New York, New York, $50-70; DuPont Chemistry Building, New York World's Fair 1939, Joy Fong Chinese American Restaurant, Brooklyn, New York, $75-100. *All courtesy of Roger Steckler.*

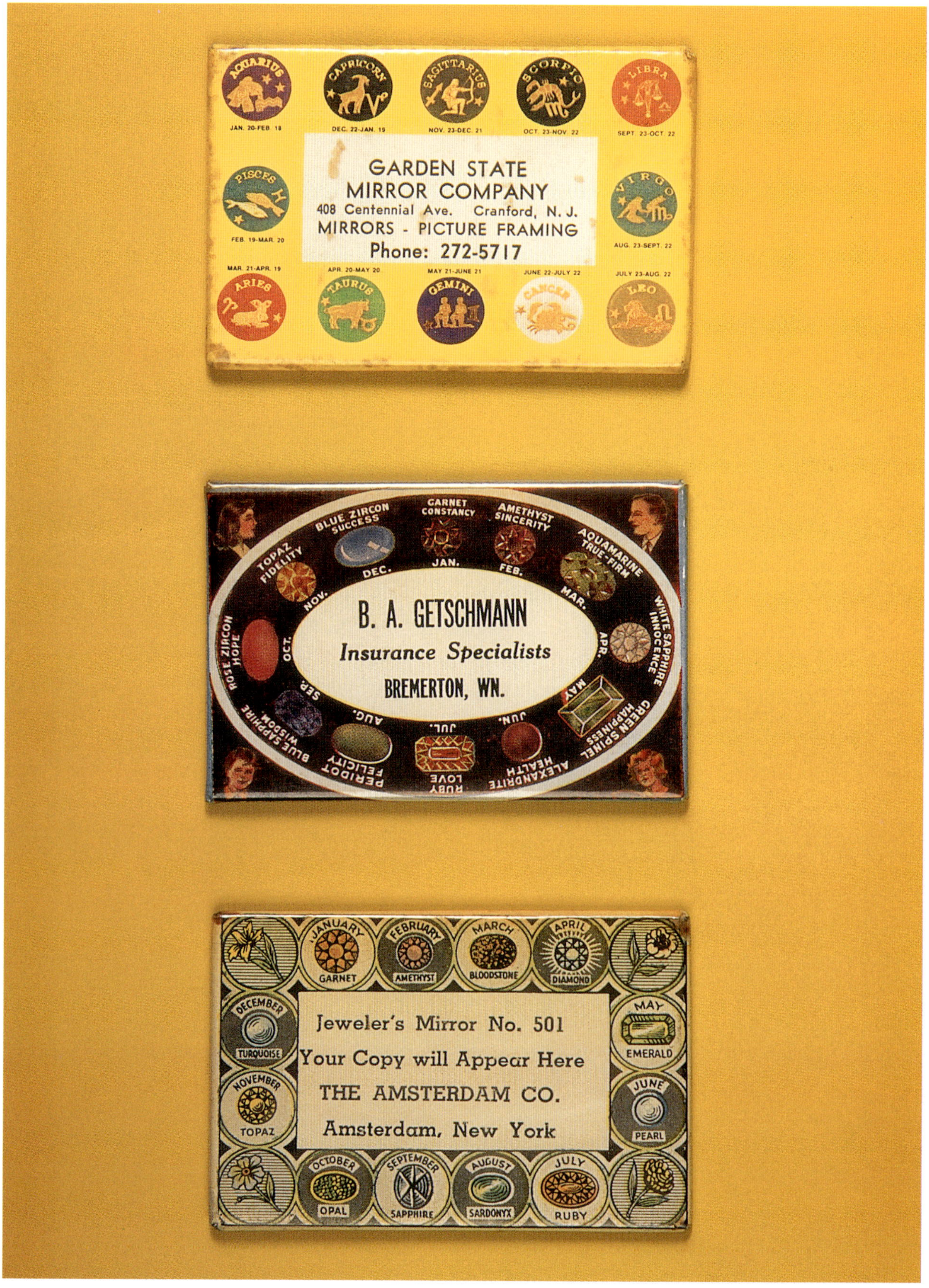

Fig. 24. *Top to bottom:* Garden State Mirror Company, Cranford, New Jersey, with zodiac symbols, $10-15; B. A. Getschmann, Insurance Specialists, Bremerton, Washington, with birthstones and their traits, $10-15; The Amsterdam Co., Amsterdam, New York, Jeweler's Mirror No. 501, with birthstones, $10-15.

Fig. 25. *Top to bottom:* Day's Ideal Lunch, Springfield, Massachusetts, \$40-50; Evelyn E. Horn, Easton, Pennsylvania (not celluloid, see Figs. 20, 21, 22), \$20-40; Madeline Beauty Parlors, Permanent Wave (New York City?), 1651 1st Ave., and 347 E. 86th St. (not celluloid, see Figs. 20, 21, 22), \$20-40.

Fig. 26. *Clockwise from top left:* C. Vernon Rettew, Special Agent, Aetna Life of Hartford, Connecticut, Harrisburg, Pennsylvania, "The Whitehead & Hoag Co., Newark, N.J." on edge, reducing mirror on reverse, $40-50; The Old Man of the Mountains, White Mountains, New Hampshire, Franconia Notch, $20-40; Massachusetts Mutual Life Insurance Co., Springfield, Massachusetts, Incorporated 1851, on fabric-like surface (see Figs. 20, 27), "Parisian Nov. Co., Chicago," on edge, $65-80; Plymouth Rock, Where Pilgrims Landed, 1620, Plymouth, Massachusetts, photo within high relief copper border, beveled mirror on reverse, $20-35.

Fig. 27. *Top to bottom, from left: Row 1,* Lydia E. Pinkham, "Yours for Health," textured fabric-like surface (see Figs. 20, 26), mirror set in from edge, to be used as container lid (see Figs. 20, 169), $40-50; *Row 2,* Dorothy Dodd Shoes, Cushing & Boucher, New Bedford (Massachusetts) (see Fig. 29), $45-75; The Invincible Junior (vacuum cleaner), Electric Renovator Mfg. Co., Pittsburgh, Pennsylvania, "Write for Free Booklets," on edge, $50-70; *Row 3,* Queen Quality Shoes (same image as on Fig. 98, full figure), $50-75.

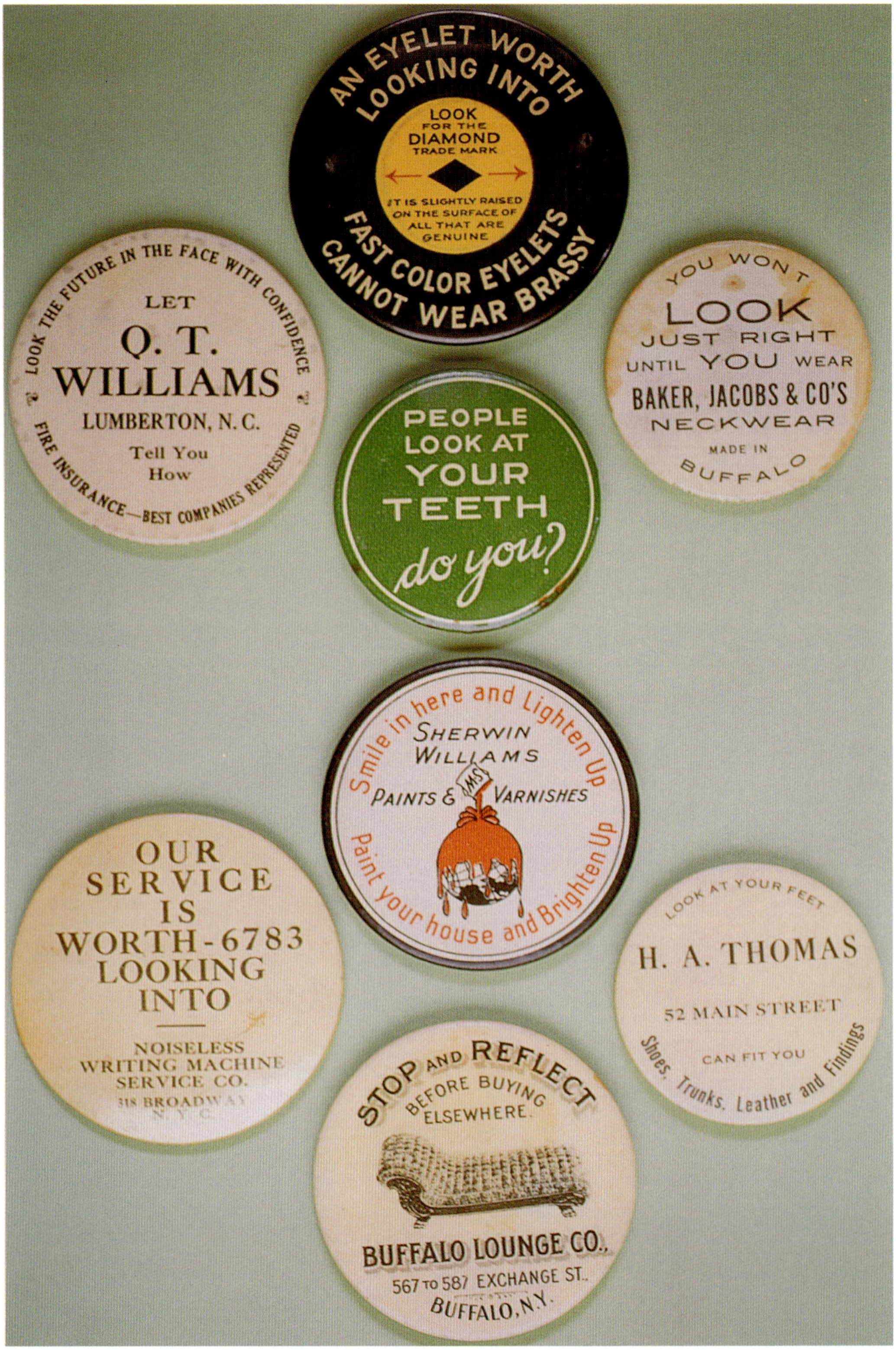

Fig. 28. *Top to bottom, from left: Row 1,* beveled mirror, "Cruver Mfg. Co., Chicago, U.S.A." on edge, $25-50; *Row 2,* Q. T. Williams, Lumberton, North Carolina, Fire Insurance, "The American Art Works, Coshocton, O" on edge, $25-50; "St. Louis Button Co., St. Louis, Mo." on edge, $25-50; Baker, Jacobs & Co.'s Neckwear, Buffalo (New York), "The American Art Works, Coshocton, O." on edge, $25-50; *Row 3,* Noiseless Writing Machine Service Co., NYC (New York), $25-50; Sherwin Williams Paints & Varnishes, "Bastian Bros. Co., Rochester, N.Y." on edge, $40-75; H. A. Thomas, Shoes, Trunks, Leather and Findings, 52 Main Street (no city, state), "The American Art Works, Coshocton, O." on edge, $25-50; *Row 4,* Buffalo Lounge Co., Buffalo, New York, "The Whitehead & Hoag Co., Newark, N.J." on edge, $50-70.

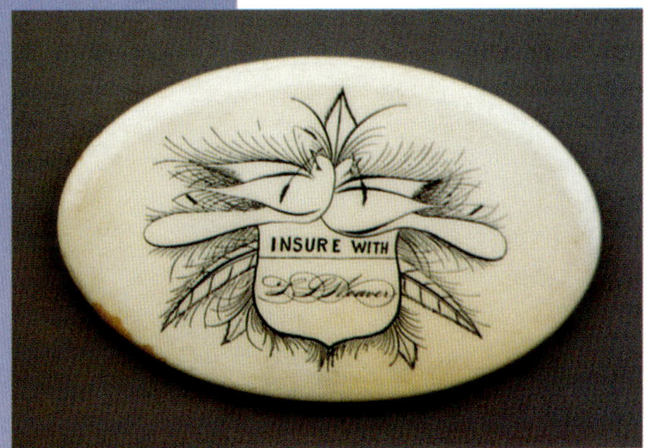

Fig. 30. "Insure with L. S. Weaver," $75-100.

Fig. 29. *Top to bottom, from left: Row 1,* Sisson Brothers, Welden Company, Binghamton, New York, "The Whitehead & Hoag Co., Newark, N.J." on edge, $25-35; *Row 2,* Cuban-American Mfg. Co., New York, Louis G. Deschler, Corina, Queen of Havana Cigars, $75-100; *Row 3,* Tea-ette, Tea, "with the (poisonous) Tannin Removed," $35-50; Hunter Tailoring Co., Cincinnati, Ohio, $35-50; *Row 4,* F. L. Hersey, Augusta, Maine, Dorothy Dodd Shoe, $35-50.

Fig. 31. *Top to bottom, from left: Row 1,* Wm. R. Mosher, Men's Furnisher, Mechanicville, New York, early all-celluloid, reducing mirror, $25-40; *Row 2,* Public Meat Market (no city, state), 47 Montgomery Street, "Baltimore Badges Novelty Co." on face, $20-30; The Plastergon Wall Board Co., Buffalo, New York, "Bastian Bros. Co., Rochester, N.Y." on edge, not a mirror, dice and bead dexterity game on reverse, $20-30; *Row 3,* Northern Telephone Company, Swanton Richford, Enosburg Falls, Vermont, $25-35; *Row 4,* The Travelers Insurance Company, Hartford, Connecticut, "The Whitehead & Hoag Co., Newark; N.J." on edge, $25-35.

Fig. 32. *Top to bottom, from left: Row 1,* Lafayette Smith Grocer Co., Cadet Brand, groceries and cigars, Springfield, Illinois, U.S.A., "The Whitehead & Hoag Co., Newark, N.J." on edge, $25-30; *Row 2,* McKesson & Robbins, New York, USA, Calox, the Oxygen Tooth Powder, $45-80; Member of Za-Ga-Zig Temple, Des Moines, Iowa, "Ask Me," name, William S. Hazard, written by hand, $20-35; *Row 3,* Krakoff's Fine Feminine Apparel, $10-25; *Row 4,* John G. Merkel Co., Coal, 20 to 28 Prince St. (no city, state; New York?), "The Whitehead & Hoag Co., Newark, N.J." on edge, $40-75; Oblinger Bros. & Makers, Vesper 5¢ Cigar, "The Whitehead & Hoag Co., Newark, N.J." on edge, $65-80.

Fig. 33. *Top to bottom:* F. M. White & Co., Druggists, "Cruver Mfg. Co., Chicago" on edge (see Fig. 28), $25-50; Bingham's Milk, "Parisian Nov. Co., Chicago," on edge, $70-100; Gerhard Lang Brewery, Buffalo (New York), Lang trademark, The Whitehead & Hoag Co., Newark, N.J." on edge, $60-75; "On Land or Sea, North, South, East or West, Stacy's Chocolates are the best," Rochester, New York, "Meet me at the Fountain, H. T. Horst, J. B. Carroll, Chicago," on edge, $35-45.

Fig. 34. *Top to bottom, from left: Row 1,* Downyflake Doughnuts, \$50-60; *Row 2,* E. S. Bodwell & Son, Brunswick (Maine), "Parisian Nov. Co., Chicago" on edge, \$35-50; Supreme Lodge Knights of Pythias Insurance Department, Indianapolis, Indiana, "The Benham-Pray Co., Indianapolis, Ind." on edge. \$40-75; *Row 3,* Reading Abattoir Co., Reading, Pennsylvania, Raco Hams and Bacon, \$35-45; *Row 4,* Raske's Flower Shop, Harrison (no state), Rotunda Board of Trade, "Parisian Nov. Co., Chicago" on edge, \$40-85; Troy Business College, Troy, New York, "The Whitehead & Hoag Co., Newark, N.J." on edge, \$30-45; *Row 5,* O. F. Jordan Company, East Chicago, Illinois, "Parisian Novelty Co., Chicago" on edge, \$30-40.

Fig. 35. *Top to bottom, from left: Row 1,* "1," "Phila. Badge Co., Phila, USA" on edge, $20-30; *Row 2,* J. W. Matthews & Co., Newburgh, New, York, Arnolds New Process Superlative Patent Flour, Trademark Registered Dec. 20, 1887, "The Whitehead & Hoag Co., Newark, N.J." on edge, $20-30; Popular Pease Pianos, New York City, New York (see Figs. 83, 123, 124), $25-35; *Row 3,* Atlantic Avenue Paving Carnival, Official Souvenir, Oct. 24, 1907 (Brooklyn, N.Y.?), framed mirror suspended from pin on double chain, $40-50.

Fig. 36. *Top to bottom:* Photomaton, rounded portrait photograph over printed background blank, in tension-held positionable holder with separate handle, "Pat. Pend.," \$30-60; Photomaton, beveled portrait photograph over printed background blank, "Philadelphia Badge Co., Phila., Pa." on edge, \$30-50; Photomaton, "Filene's," rounded portrait photograph over printed background blank, "Philadelphia Badge Co., Phila, Pa. on edge, \$30-50; Photograph under open paper edging, Jerome De Mott, Amsterdam Avenue at 81st St. (New York City, New York), \$30-50.

Photo Machine Portraits & a Collage

Fig. 37. *Top to bottom:* Woman in cloche and pearls, tinted photograph on textured gold surface, c. 1930s, $30-40; Cut-out portrait on silver foil, c. 1930s, $30-40; Boy with collar (mother's hand holding arm) on gold-toned ground, c. 1930s, $30-40.

Child's Portrait

Fig. 38. Child in ceremonial garb, undated, Cuba (possibly a memorial token), $30-60.

Fig. 39. *Top to bottom, from left: Row 1,* Sheila Stark Brummel, 1935, New York, New York (with hand of mother, Rose Stark). *Courtesy of Sheila Brummel; Row 2,* little boy, nude, tinted cheeks and lips, $10-30; girl with flower, $10-30; *Row 3,* John Todaro in knitted hat and sweater, 1939, New York City, New York. *Courtesy of John Todaro.*

Fig. 40. *Top to bottom, from left: Row 1,* Bar Mitzvah, boy with prayer shawl, c. 1940s, $15-25; *Row 2,* First Communion, girl with garland, bouquet, and bible, c. 1910s-20s, $15-25; First Communion, boy with silk bow and bible, c. 1910s-20s, $15-25; Confirmation, girl with candle and head veil, c. 1930s-40s, $15-25; *Row 3,* First Communion, John Todaro, tinted photograph, New York City, 1944. *Courtesy of John Todaro*; First Communion, girl with ribbon and veil, c. 1940s, $15-25.

Fig. 41. *Top to bottom, from left: Row 1,* Fireman with cap, c. 1940s, $15-30; Graduate in cap and gown, c. 1940s-50s, $15-30; *Row 2,* Victorian gentleman in derby and glasses, $15-35; *Row 3,* Boy in sailor suit (tinted background), c. 1900s-10s, $20-35; Mother and child (in sailor suit), tinted photograph, c. 1930s, $20-35.

Fig. 43. Miss Ulrich of Warsaw, c. 1933. Not a mirror, this 5-1/2 x 4-1/4 tinted photographic portrait, enameled on metal with paper backing, is typical of many reduced in size for reproduction as pocket mirrors from the late nineteenth to mid-twentieth century. *Courtesy of Miss Ulrich's great-niece, Juliette Galant.*

Fig. 42. *Top to bottom, from left: Row 1,* The Matriarch, tinted photograph, turn of the century dress, c. 1890s, $20-30; Snapshot, outdoors, with dog, c. 1930s-40s, $10-15; *Row 2,* Myra Kolber Gassner, Chicago, Illinois, 1937. *Courtesy of her granddaughter, Shifra Gassner*; *Row 3,* Young girl (confirmation?), c. 1930s, $20-30; Woman in cloche and furs, c. 1920s, $20-30.

Fig. 44. *Clockwise from top:* Bride with bouquet, tinted background, Jeanne Chess Levin, January 2, 1936, Brooklyn, New York. *Courtesy of her daughter, Marilyn Cuff;* Woman with draped cap and pendant, beveled mirror, c. 1930s, $20-30; Woman with corsage and pearls, c. 1930s, $20-30; Woman with pearls and earrings, tinted photograph, c. 1920s-30s, $20-30.

Fig. 45. *Top to bottom, from left: Row 1,* Bride, tinted photo inset in stock opalescent frame, LerMan Studio . . . , Passaic, New Jersey, c. 1920s-30s, \$20-30; *Row 2,* Bride, "Aunt Mary" Vacasi Castrovinci, tinted photograph, New York City, c. 1950s, "Portrait by A. Ruisi, N.Y." on edge. *Courtesy of her nephew, John Todaro*; Bride in studio, tinted photograph, c. 1940s, \$20-30; *Row 3,* Bride, tinted photograph, c. 1940s, \$20-30; Bride, with orchid headdress, c. 1940s, \$20-30.

Brides & Grooms, Rectangular

Fig. 46. *Top to bottom, from left: Row 1,* Claudita Castrovinci and Joseph Todaro, 1937, New York City, New York, tinted photograph. *Courtesy of their son, John Todaro*; Bride and groom, c. late 1920-30s, $25-30; *Row 2,* Bride and groom, c. 1930s, $25-30; *Row 3,* Bride and groom, in uniform, tinted photograph, c. 1940s, $25-30; Lucia Maria Dolci and Josef Rudolph Rindler, New York City, 1940, Parents of Robert Rindler, Dean, Cooper Union School of Art. *Courtesy of Dean Rindler.*

Brides & Grooms, Diagonal Ovals

Fig. 47. *Top to bottom, from left: Row 1,* Bride & groom, tinted photograph, c. 1920s-30s, "Santoro Studio . . . , Brooklyn, New York," $25-40; Bride and groom, tinted photograph, c. 1940s, $20-30; *Row 2,* Bride and groom, tinted photograph, c. 1940s, $20-35; Bride and groom, tinted photograph, c. 1910s-20s, $25-40; *Row 3,* Bride and groom, tinted photograph, c. 1930s, $20-35.

Brides & Grooms, Tinted

Fig. 48. *Top to bottom, from left: Row 1,* Bride and groom, tinted photograph, c. 1940s-50s, $15-25; Hortense Rosenberg and Julius Herlitschek, tinted photograph by Rappaport at Café Loyale, New York City, March 30, 1941. *Courtesy of their daughter, Florence Baravarian*; *Row 2,* Bride and groom, tinted photograph, c. 1930s, $25-35; *Row 3,* Bride and groom, tinted photograph, c. 1930s-40s, $20-35; Bride and groom, tinted photograph, c. 1930s-40s, $20-35.

Fig. 49. *Clockwise from upper right:* Gemma Margaret Gambone and Ralph Giocolano, Our Lady of Pompeii, Roman Catholic Church, Carmine Street, New York City, 1933. *Courtesy of their daughter, Concetta Crespo;* Bride and groom, tinted photograph, c. 1920s-30s, $25-35; Bride and groom, c. 1920s-30s, $20-30; Bride and groom, tinted photograph, c. 1920s-30s, $25-35; Bride and groom, tinted photograph, c. 1920s-30s, $25-35.

Fig. 50. *Top to bottom, from left: Row 1,* Bride and groom, tinted photograph, c. 1930s, $25-35; Bride and groom, tinted photograph, c. 1920s, $25-35; *Row 2,* Bride and groom, tinted photograph, c. 1920s, "Strauss Studio, A. Kagan, President, 239 South 4th St., Brooklyn, New York" along border, $25-40; *Row 3,* Bride and groom, tinted photograph, c. 1920s, $25-40.

Fig. 51. *Top to bottom:* Bride with bouquet, tinted background, c. 1920s-30s, $30-40; Full length bride and groom, tinted photograph, c. 1930s, $20-30; Standing bride and groom, tinted photograph, c. 1950s, $15-25.

Gilt-Edged Studio Portraits

Fig. 52. *Top to bottom, from left: Row 1,* First Communion portrait, father and son, c. 1930s, $25-40; Bride, Ramon Photo Studios, Brooklyn, New York, c. 1920s-30s, $20-35; *Row 2,* Man in uniform, "Barresi Art Studios, Bklyn., N.Y." on edge, c. 1940s, $25-35; *Row 3,* Bride and groom (in top hat), Goldsmith Studio, Brooklyn, New York, c. late 1920s-30s, $35-45; Bride and groom, The Morgen Studios, Brooklyn, New York, c. 1930s, $25-35.

Fig. 53. *Top to bottom, from left: Row 1,* Child in knitted cap and bows, tinted photograph, Farb's Photo Studio, 156 Tompkins Avenue, Brooklyn, New York, c. 1940s. *Courtesy of Christina and Irwin Rubin; Row 2,* Bride with cross and scroll, tinted photograph, Hollywood Art Studio, 105-10 Crossbay Boulevard, Ozone Park, New York, c. 1930s-40s, \$25-40; Bride and groom, tinted photograph, The Morgen Studios, 5101 New Utrecht Avenue, Brooklyn, New York, c. 1930s-40s, \$25-40; *Row 3,* Bride and groom, tinted photograph, Damoff Studio, 37 Monroe Street, Passaic, New Jersey, c. 1930s-40s, \$25-40.

Niagara Falls

Fig. 54. *Top to bottom, from left: Row 1,* Niagara Falls, tinted artwork, "Bastian Bros., Rochester, N.Y." on edge, $40-50; Souvenir, Niagara Falls, tinted photograph, $25-40; *Row 2,* Niagara Falls, tinted photograph on textured paper under celluloid, "Louis Bernheimer Co., Mfr., 2077 E. 4th St., Cleveland, Ohio" on edge, $15-30; *Row 3,* Souvenir, Niagara Falls, tinted photograph, $25-45; Souvenir, Niagara Falls, tinted photograph, $25-35.

Fig. 55. *Top to bottom, from left: Row 1,* Silver Wedding Anniversary, on silver double frame with beveled mirror, separate tinted photographs, c. 1920s-30s, $25-35; *Row 2,* Off-center stock floral framed tinted portrait, c. 1930s, $30-45; Opalescent patch stock frame, bride with white hat, c. 1930s, $30-45 *Row 3,* Floral stock framed bearded man's portrait, slightly off-center, c. 1920s-30s, $30-45; *Row 4,* Off-center stock floral frame with woman's tinted portrait, c. 1920s-30s, Pennsylvania, $30-45; Off-center stock floral frame with woman's tinted portrait, c. 1920s-30s, $25-35.

Family Groupings

Fig. 56. *Clockwise from upper right:* Two women (mother and daughter?), Will Art Studio, Brooklyn, New York, c. 1920s-30s, $15-25; Mr. & Mrs. Morris Elster, Golden Wedding Anniversary, December 20, 1931, $25-35; Elderly couple on house steps, location unknown, c. 1910s, $15-25; Family, individual portraits (mother, father, son, daughter), c. 1930s, $25-35; Two women (mother and daughter?), tinted photograph, c. 1930s, $35-45.

My Summer Vacation . . .

Fig. 57. *Clockwise from upper left:* Mother and infant ("apartment to lease 4 rooms" in background), c. 1920s-30s, $15-25; Girl and infant in backyard garden, c. 1920s-30s, $10-15; Young women at the beach, c. 1930s, $25-30; Young men at the mountains, c. 1910s-30s, $15-20; Woman reading in the countryside, c. 1930s, $10-20.

Fig. 59. Youth with pensive look, $40-50. *Courtesy of Roger Steckler.*

Fig. 58. *Top to bottom, from left: Row 1,* Woman outdoors, with shawl, c. 1930s, $10-20; *Row 2,* Woman, among trees, c. 1930s, $10-20; Same woman as second mirror, arms on waist, c. 1930s, $10-20; *Row 3,* Young man in jacket and tie, c. 1930s-40s, $15-25.

Fig. 60. *Top to bottom, from left: Row 1,* Standing man in suit and tie, tinted photograph, c. 1920s-30s, $20-30; Man with pompadour, with wing collar, tinted photograph, c. 1910s-20s, $15-25; *Row 2,* Young man with glasses and widespread ears, tinted photograph, c. 1910s-30s, $20-30; *Row 3,* Regal man in suit, with ribboned boutonniere (groom?), tinted photograph, c. 1930s, $25-35; Man with bow tie and boutonniere, c. 1920s, $15-25.

Fig. 61. *Clockwise from upper left:* M. Dorn, Burlington, Vermont, "The Meek Co., Coshocton, O." on edge, $50-65; Wm. M. Leighton, Tailor, Importer, Portland, Maine, $50-60; *Herrmann the Great*, Direction of Ed. Thurnaer (see Fig. 62), $250-400; J. W. Whyte, Prop. Enterprise Custom Laundry, Lynn, Massachusetts, "A. R. Lopez" on edge, $65-80; Mahlon N. Haines, The Shoe Wizard, "Parisian Nov. Co., Chicago" on edge, $45-65.

Early Theatrical . . .

Fig. 62. *Clockwise from upper left:* Fay Marbe (signature), Belmont Theater, $35-50; Miss Valerie Valaire, Chicago Stock Company, Chas. H. Rosskam, Mgr., $40-50; Mae Marsh, "Hartsook Photo, S. F., — L. A." on image (see Fig. 63), $50-65; Francis Ford, Grace Cunard, *The Broken Coin*, Universal, 15 weeks, 30 reels, "The Whitehead & Hoag Co., Newark, N. J." on edge, $65-80; Sherman S. Krellberg offers Helen Holmes in *The Fatal Fortune*, $40-55.

. . . and a Sweetheart!

Fig. 63. *Top to bottom:* Mae Marsh, "Hartsook Photo, S.F./L.A." on face (see Fig. 62), $50-65; Mary Pickford, "Hartsook Photo, S.F./L.A." on face, $100-150.

Groups

Fig. 64. *Top to bottom:* Westfield Theater, First Anniversary, 1912-1913, $25-30; Helen May Butler and her Ladies Brass Band, "Willow Grove Park (PA?)," on edge, $40-50; Jasper Basket Ball Team, Champions, 1913-1914, "Bastian Bros. Co., Rochester, N. Y." on edge, $35-45.

Fig. 65. The Doll Family: Grace, Harry, Daisy & Tiny, $200-300. *Courtesy of Roger Steckler.*

Fig. 67. Louis Jordan, with signature, $45-65. *Courtesy of Roger Steckler.*

Fig. 66. *Top to bottom:* Count Basie, with signature, $45-65; Dionne Quintuplets, ©N.E.A. (with brown velvet muff), $40-60; Dionne Quintuplets in basket, ©1936 by N.E.A. Service, Inc., $40-60.

Fig. 68. *Top to bottom, from left: Row 1, Don't Bother to Knock*, Richard Widmark and Marilyn Monroe, 20th Century Fox, $25-50; *Casablanca*, Humphrey Bogart, Ingrid Bergman, Paul Henreid, Warner Bros. (in French), $30-50; *Row 2,* "Marilyn Monroe, in *Gentlemen Prefer Blondes*" and "©1953 20th Century Fox Film Corp." on edge, $20-30; Rita Hayworth look-alike (signed lower left by artist), Vargas-like inscription. $20-30.

Fig. 69. *Top to bottom:* "Elvis and his parents," 1956, $5-15; Elvis Presley with guitar, no inscription, color photograph, new, $10-30.

Fig. 70. *Top to bottom:* "In the waist-coat-pocket, dice box with dice in metal," with five metal dice, \$80-90; "Lewis 66" Rye, Strauss, Pritz & Co., Distillers, Cincinnati, Ohio ("made in Germany"), with five dice, \$80-90; The International Tailoring Co., Chicago, New York, Berlin, London, with image of lion, five-dice collection, \$70-85. *Courtesy of Christina and Irwin Rubin*; Chas. H. Childs & Co., Utica, New York, "Carriages, implements, harness" (basket hangs from balloon's pin), "D.R.G.M., 156021" on face, \$125-200.

Dexterity White Bead Puzzles

Fig. 71. *Clockwise, from top:* Soccer players, beads as balls (Germany), $30-40; Woman of color with earrings, beads as eyes and teeth, $50-60. *Courtesy of Christina and Irwin Rubin;* Rooster atop basket, beads as eggs, "D.R.G.M. 137321" (Germany), $80-100; The International Tailoring Co., New York, Chicago, Berlin, London, "King of Tailors" with Lion, beads as teeth, $75-90; Monkeys at billiard table, beads as balls, "D.R.G.M. 116769" (Germany), $130-145. *Courtesy of Christina and Irwin Rubin.*

Oh, Those Shifty Eyes!

Fig. 72. Bald gentleman with shifting (but contained) black bead eyes, white bead teeth, embossed relief image, "D.R.G.M. 180082" on inner edge, $200-250.

Fig. 73. *Top to bottom:* Tennis match (Germany), $30-40. *Courtesy of Christina and Irwin Rubin;* Palm trees with street light globes (Germany), $40-50. *Courtesy of Christina and Irwin Rubin;* Bow and arrow with lion's tail at beach, $60-75. *Courtesy of Christina and Irwin Rubin;* Cow with large red eyes, "Made in U. S. zone Germany" (see Fig. 78), $40-50; Tennis match, men and women, no mirror on reverse, $70-85.

Fig. 74. *Top to bottom:* Sunshine Bread (three beads in balloons), all-metal with clear plastic cover (not a mirror), $20-30; Cracker Jack, Rueckheim Bros. & Eckstein, Chicago, Illinois (Germany) (black beads in holes), $55-85; Black cars into underpass in city, "EPESO, Germany, CE," $40-50. *Courtesy of Christina and Irwin Rubin;* Three birds, black beads as eyes, $30-45; Dachshund with cats, beads as eyes, $35-50. *Courtesy of Christina and Irwin Rubin.*

Fig. 76. White bead, dexterity gambling toy, "Germany, A Souvenir from the Universal Theatres Concession Co., Chicago, Ill." embossed on wide metal edge, $40-60.

Fig. 75. *Top to bottom:* Sports spinner, place spinner in depression, eight sports depicted, $50-85; Dice toss, shake three balls into spaces for each, "D. R. G. M., Germany," $50-85; Roulette motif, shake ball into numbered metal cup, $40-65; Soccer field, place ball into goal area, "Made in Germany," $40-65. *All courtesy of Christina and Irwin Rubin.*

Fig. 78. Cat with magnetic mouse, to be guided into metal "trap," "Made in U.S. Zone, Germany. D.R.G.M." on face (see Fig. 73), $20-50.

Fig. 77. *Top to bottom:* Woman's high shoe with black bead buttons (not a mirror, reverse side shown in Fig. 10, Row 2), $25-35; Harold Lloyd, eyeglasses to be placed on face, "DRGM, made in Germany, patent date 8/12/25," dexterity toy, no mirror on reverse, $100-200. *Courtesy of Christina and Irwin Rubin;* Elephant catching two balls in cup held by trunk, "D.R.G.M.," dexterity toy, no mirror on reverse (stamped "Germany" on reverse), $70-85. *Courtesy of Christina and Irwin Rubin;* Magnetic mouse, to be guided into metal "trap" in butcher shop. "D. R. P. No. 149257," $85-110. *Courtesy of Christina and Irwin Rubin.*

Strip Tease Flip-Flops

Fig. 79. *Top to bottom:* Male flamenco dancer, right, Spanish drapery-clad female dancer, left, "Made in Germany D.R.G.M.," $35-50; Elderly gent sleeping in chair, right, fully dressed woman in hat and furs, left, "Made in Germany D.R.G.M.," $35-50; Sad Sultan, seated, right, dancer with many veils, left, "Made in Germany D.R.G.M.," $35-50. All are enamel on metal panels.

Fig. 80. The same strip tease flip-flops, when flipped over at central hinge. *Top to bottom:* Matador, right, Spanish female, fully dressed, left; Alert elderly gent seated, right, scantily clad female, left; Gesturing Sultan, right, semi-nude dancer with veil, left.

Fig. 81. *Top to bottom, from left: Row 1,* Woman, hand to throat, "naughty" when inverted, $40-50; *Row 2,* variants of left and right, $35-45; *Row 3,* Woman, gloves to face, "anatomical" view when inverted (in color), $50-65; *Row 4*, another woman with gloves to face (sepia), turn of century, $100-200; *Row 5,* Woman with necklace, "anatomical" view when inverted (sepia), left and right, $50-75.

Fig. 82. *Clockwise from upper left:* Kleinert's Dress Shields, "Bastian Bros. Co., Rochester, N.Y." on edge, $30-40; Cooper Underwear Co., Kenosha, Wisconsin, White Cat Union Suits, "Parisian Novelty Co., Chicago" on edge, $40-65; E. C. Snow, Fairport, New York, Klosed-Krotch Union Suits, White Cat Brand, "Parisian Nov. Co., Chicago" on edge, $55-80; Worcester Elastic Stocking & Truss Co., Worcester, Massachusetts, "Cruver Mfg. Co., Chicago, U.S.A., 3c" on edge, $75-85; Kleinert's Dress Shield Guimpe, $30-40.

Fig. 83. *Top to bottom, from left: Row 1,* Smith's Green Mountain Renovator, "The Whitehead & Hoag co., Newark, N.J." on edge, $75-85; Hall, Warren, Ohio, Pianos, Victrolas, Sewing Machines, "Geiger Bros., Newark, N.J." on edge, $60-75; *Row 2,* Bailey's Music Rooms, Lancaster, New Hampshire, Pianos, Inner-Player Pianos (see Figs. 16, 123, 124), $45-55; Huguenin Medailleurs, LeLocle, embossed brass relief of painter at easel signed l.r., "N. Huguenin," beveled mirror, French, $35-45; *Row 3,* The Sharples Separator Co., West Chester, Pennsylvania, Chicago, Illinois, Toronto, Canada, $100-150; Ridley & Flanders, Oakland, Maine, The Elite Petticoat, patented Jan. 19, 1904, "Jackson Skirt & Novelty Co., Jackson, Mich." on edge, $45-60.

Holidays, Smiles, Familiar Figures

Fig. 84. *Clockwise from upper left:* Performer, mask and hat, Topeka's Second Annual Carnival, KK, "Pat. July 21, 1896" on edge, $45-65; Morton's Salt, "Lou Fox, Chicago" on edge, $30-40; (Santa), Guardian Trust Company of New Jersey, "Landis Savings Club Co., Harrisburg, Pa." on edge, $60-85; "I Buy my Harness of" Wilmer A. Young, Winchester, New Hampshire (smiling face), frowning face when inverted, "The American Art Works, Coshocton, O." on edge (see also Fig. 81), $60-90; Cracker Jack (New!), "Reg. U.S. Pat. off." (see Fig. 74), $5-10; (Holly and Berries), Diamond Drug Store, East Liverpool, Ohio (union label: "Allied Printing Trades Council, Newark, N.J., 3"), "The Whitehead & Hoag Co., Newark N.J." on edge, $35-45.

Fig. 85. *Top to bottom, from left: Row 1,* Mennen's Flesh Tint Talcum, "Parisian Nov. Co., Chicago" on edge, $50-75; Mennen's Violet Talcum Toilet Powder, "Duplicate Mirror 5¢ postage: Gerhard Mennen Co., Newark, N.J." on edge, $45-70; *Row 2,* Mennen's Cream Tint Talcum, Gerhard Mennen Chemical Co., Newark, New Jersey, U.S.A., "Parisian Nov. Co., Chicago" on edge, $30-50; *Row 3,* John W. Crooks Chocolate Company, Boston, Massachusetts, "The American Art Works, Coshocton, O." on edge (see Fig. 29), $40-50; Rosary Chocolates, Mueller-Keller Candy Co., St. Joseph, Missouri, "Parisian Nov. Co., Chicago" on edge, $45-60.

Fig. 86. Stock image: Mennen's Borated Talcum Toilet Powder, "Duplicate mirror 5¢ postage Gerhard Mennen Co., Newark, N.J." on edge, $100-150.

Fig. 88. Stock image, used again! Charles H. Netsch, The Artistic Florist, Buffalo, N.Y., "Union Label, Allied Printing Trades Council, Rochester," very small, on face. $150-200.

Fig. 87. *Top to bottom, from left: Row 1,* Wm. Dethlefs, Florist, Mitchell, South Dakota, "The Coshocton Specialty Co., Coshocton, Ohio" on edge, $40-50; *Row 2,* Dancing every Friday night at Doelger's, Prospect Park Hall, 358 Morris Avenue (no city, state) (NYC?), "The Whitehead & Hoag Co., Newark, N.J., copyright 1903" on edge, $45-60; *Row 3,* LaFrance Shoe for women, sold by: Gottleib's Dept. Store, Point Pleasant, New Jersey, "J. B. Carroll, Chicago" on edge (with reducing mirror on reverse), $50-65; LaFrance Shoe for Women, sold by: Preble & Robinson, Bingham, Maine, "J. B. Carroll, Chicago" on edge (with reducing mirror on reverse), $50-65; *Row 4,* Helen·Style Shop, Van Hook, North Dakota, No. 539, $30-40.

Fig. 89. *Counter-clockwise, from upper right:* Alvin Theater, Mansfield, Ohio, Prof. H. Paul Bertholy, 1919, "Sample No. 1237" on face, all celluloid folding handle type, $35-50; Hyatt Mfg. Co., Baltimore, Maryland, Badges and Buttons of Every Kind, from San Antonio Reunion, Texas Cow Punchers, $80-100; Ladies Night, Reading Consistory, June 21, 1922, all celluloid folding handle type, $25-35; Lewiston and Auburn (Maine), Shrine Club, Colleen Cabaret, Mar. 13, 1922, folding celluloid handle, $30-50; Rothschild Bros. & Co., Buttons Exclusively, New York, established 1850, $85-100.

Mostly Good For . . .

Fig. 90. *Top to bottom, from left: Row 1*, Hotel Raymond, Fitchburg, Massachusetts, tinted photo, "Cruver Mfg. Co., Chicago, H.P.P." on edge (woman in profile, flower in hair), $100-200; Seneca Hotel, Baldwinsville, New York, Portrait of Evelyn Nesbit, "Cruver Mfg. Co., Chicago" on edge, $150-200; *Row 2,* Hotel Raymond, Fitchburg, Massachusetts, "Cruver Mfg. Co., Chicago, H.P.P." on edge (woman in middy blouse, red coloring on hat), $125-150; Hotel Raymond, Fitchburg, Massachusetts, "Cruver Mfg. Co., Chicago, H.P.P." on edge (two dogs, green coloring on grass), $80-100; *Row 3,* Commercial Club, Falls City, Nebraska, "Cruver Mfg. Co., Chicago, H.P.P." on edge (child with finger in mouth, tinted background), $150-200; N. Greenberg's, Lewiston, Maine, "Cruver Mfg. Co., Chicago" on edge, $50-75.

Two Rarities

Fig. 91. Good for 10¢ in trade, The Corner, Middletown, Connecticut, reclining semi-nude in pink with flowers, Cruver Mfg. Co., Chicago on edge, $450-500. *Courtesy of Mark and Lynda Suozzi, Ashfield, Massachusetts.*

Fig. 92. Good for 10¢ in trade, The Corner, Middletown, Connecticut, seated semi-nude in yellow garment, Cruver Mfg. Co., Chicago on edge, $450-500. *Courtesy of Mark and Lynda Suozzi, Ashfield, Massachusetts.*

Fig. 93. *Top to bottom:* "I use Red Cross Shoe Polishes," copyrighted, "F. F. Pulver Co., Rochester, N.Y." on edge, $40-60; Isabelle Underwood, Souvenir of Pleasure Bay, Season 1902, illegible inscription on edge, possibly ". . . Bangor . . . " (see Fig. 62), $40-60; A Pentucket Rollway Girl, a "Good for" (see Fig. 90), $200-300.

Fig. 94. *Top to bottom, from left: Row 1,* Souvenir of Des Moines, Iowa, $30-45; *Row 2,* C. F. Bilodeau - Clothier, Augusta (Maine), woman with plumed hat, $100-200; C. F. Bilodeau - Clothier, Augusta (Maine), woman with choker and lace, $100-200.

Classic Ladies

In Spanish

Fig. 95. Cigarros Trinidad y Hro, No. 223 (Cuba) (see Fig. 108), $50-75.

Fig. 96. *Clockwise from top, and center:* Edwin J. Cummings, High Class Coal, Philadelphia (Pennsylvania), art by Angelo Asti, "Copyright 1905, The Whitehead & Hoag Co." on edge, $200-250; Oxford Chocolates (woman in profile, in cap and gown), "Duplicate mirror by mail for 5¢ postage, Hazen Confectionery Company" on edge, $150-175; Woman with red flower in long hair, "Eleanor," art by Angelo Asti, no inscription, $60-80; Oxford Chocolates (woman, front view, in cap and gown), "Duplicate mirror by mail for 5 cents postage, Hazen Confectionery Co., Boston, Mass." on edge, $150-175; *Center:* Carmen Complexion Powder, Stafford Miller, Co., St. Louis, Missouri, $60-80.

Fig. 97. *Top to bottom, from left: Row 1,* Tainter's, Lewiston (Maine), Music, Pianos (girl with head scarf), $50-175; Tainter's, Lewiston (Maine), Music, Pianos (girl with cat), $200-250; *Row 2,* Chichester Chemical Co., Philadelphia (Pennsylvania), The Whitehead & Hoag Co., Newark, New Jersey, $150-175; *Row 3,* Rawling's Implement Co., Baltimore, M.D. (sic), $80-125; Hirsch Bros. (no location), $50-75; *Row 4,* The Paris Millinery Shop, Poughkeepsie, New York, "Parisian Nov. Co., Chicago" on edge, $200-300; *Row 5,* C. F. Graser's Sons, Caterers, Troy, New York, $200-300; Olympia Shoe, Hat Cleaning Parlor, 128 North Queen St. (no city, state), "Made in America, Parisian Nov. Co., Chicago," $200-300.

Fig. 98. *Clockwise from top, and center:* Queen Quality Shoes, "J. G. Harding, Sole Agent, Middletown, N.Y." on edge, $100-125; "The Girl of the Pingree Shoe," "Copyrighted The Pingree Co. Detroit, 1910, Bastian Bros. Co., Roch. N.Y." on edge, $150-200; The May Company, art by Angelo Asti, "Copyright 1906 by the Whitehead & Hoag Co." on edge, $80-110; Holden's, The House of Glass, $75-80; Noonan Bros. Home Furnishing, Mahanoy City, Pennsylvania, art by Angelo Asti, "The American Art Company . . ." on edge, $80-125; Oxford Chocolates, art by Angelo Asti, "Hazen Confectionery Co., Boston, Mass." on edge, $100-125; *Center:* Huyler's, art by Angelo Asti, "The Whitehead & Hoag Co., copyright 1903, Newark, New Jersey" on edge, $80-125.

Fig. 99. *Top to bottom:* Massachusetts Corset Co., Wilhelmina Corsets, "The Whitehead & Hoag Co., Newark, N.J." on edge, $70-85; Norse-American Centennial, Haakon VII, King of Norway, Queen Maud – Crown Prince Olav, 1925-1925, Minneapolis-St. Paul (Minnesota), "Cruver Mfg. Co., Chicago" on edge, $125-150; Sparrow's Empress Chocolates, "Parisian Nov. Co., Chicago" on edge, $60-85.

Fig. 100. *Top to bottom, from left: Row 1,* Dold Poultry Feed, "Bastian Bros., Rochester, NY" on edge, $675; Ondoca, "Made in Canada" on edge, $500; Adams' Pepsin Tutti Frutti Gum, "Whitehead & Hoag, Newark, NJ" on edge, $850; *Row 2,* Omar Pearls, "Bastian Bros, Balt, MD" on edge, $700; R. R. Overalls, $3,000; Surbrug Tobacco, "Send 6 cents Surbrug, 132 Reade St., NY for duplicate" on edge, $650; *Row 3,* Burrows Billiard Table, "Cruver Mgr Co., Chicago, Ill" on edge, $800; Progress Cooking Oil, "Whitehead & Hoag, Newark, NJ" on edge, $800. *All courtesy of Burt Purmell.*

Fig. 101. *Clockwise, from left:* Keith's Theaters, "Whitehead" or "Bastian" underneath stretched celluloid(?), $750; Gibson Furniture, "Bastian Bros, Rochester, NY" on edge, $250; Egg Baking Powder, "Baltimore Badge & Novelty Co. Balt, MD" on edge, $750; Franklin Mills Co., "Bastian Bros, Rochester, NY" on edge, $475. *All courtesy of Burt Purmell.*

Fig. 102. *Clockwise, from left:* Bosch Spark Plugs, no manufacturer name, brass collet around celluloid, $800; Columbia Flour, "Bastian Bros, Rochester, NY" on edge, $750; Ryan's Pure Beers, $550. *All courtesy of Burt Purmell.*

Burt's Beauties

Fig. 103. *From left:* Keystone Overalls, "Send 6 cents for duplicate, Keystone Working Garments for 25 years the best, Whitehead & Hoag, Newark, NJ" on edge, $1,500; Hires Root Beer, "For duplicate mirrors send 5 cents in stamps to Chas Hires, Phila., Whitehead & Hoag, Newark, NJ" on edge, $1,500. *Both courtesy of Burt Purmell.*

Schuman's Specialties

Opposite page:
Fig. 104. *Top to bottom, from left: Row 1,* "Limetta," Drewry & Sons, St. Paul, Minnesota, $1,000-1,500; Old Manse Canadian Sap Maple Syrup, William R. Manierre, $1,500-2,000; Cream Dove Shortening and Peanut Butter, Cream Dove Mfg. Co., Binghamton, New York, $750-1,000; *Row 2,* The Call Transfer Co., Auto Service, 1301 J. St. (no city, state). $500-750; Egg Baking Powder (white hat), $1,000-1,500; Egg Baking Powder (red hat), $1,000-1,500; *Row 3,* A. R. Rugaber, Furniture and Floor Coverings, Westfield, Pennsylvania, $500-750; Knox's Gelatine Running Team, Johnstown, New York, Compliments of Tryon Hook & Ladder Co., $500-750; The Providence Car Fender, The Consolidated Car Fender Co., Providence, Rhode Island, $750-1,000; *Row 4,* Ideal or Progress Cooking Oil, Louisville, Kentucky, $750-1,000; The Berry Brothers' Toy Wagon, Berry Brothers Varnishes & Architectural Finishes, $500-750; Buster Brown Shoes, The Brown Shoe Co., St. Paul's, Missouri, $750-1,000. These mirrors were mostly made by the Whitehead & Hoag Company, Newark, New Jersey, c.1900-1910s. *All courtesy of Robert Schuman, M.D.*

"LIMETTA"
THE DRINK OF DRINKS
ORDER A CASE FOR YOUR HOME
NOTICE TRY LIMETTA HIGH-BALL
PUT UP ONLY BY DREWRY & SONS, ST. PAUL.
"OLD MANSE"
CANADIAN MAPLE SAP SYRUP
IS ABSOLUTELY PURE
"IT'S LICKIN' GOOD"
IMPORTED & CANNED ONLY BY
WILLIAM R. MANIERRE.
LOOK INTO CREAM DOVE SHORTENING AND CREAM DOVE BRAND PEANUT BUTTER
CREAM DOVE MFG. CO., BINGHAMTON, N.Y.
WHY LOOK FOR A WAGON, WHEN WE CAN AN AUTO
THE CALL TRANSFER CO.
O.E. FULLER, Prop. 1301 J ST. Always on the Job. Prices Right
AUTO SERVICE Phone 1900-R
EGG BAKING POWDER
EGG BAKING POWDER
A. R. RUGABER
FURNITURE and FLOOR COVERINGS
Cash or Credit
SPECIAL PRICES TO NEWLYWEDS
WESTFIELD, PA.
KNOX'S GELATINE RUNNING TEAM
JOHNSTOWN, N.Y.
COMPLIMENTS OF
TRYON HOOK & LADDER CO.
THE PROVIDENCE CAR FENDER
SAVES HUMAN LIVES
THE CONSOLIDATED CAR FENDER CO., PROVIDENCE, R.I.
IDEAL OR PROGRESS COOKING OIL
A PURE, HEALTHFUL, VEGETABLE OIL FOR HOME USE.
BERRY BROTHERS VARNISHES
THE BERRY BROTHERS' TOY WAGON
TAKEN FROM LIFE.
BUSTER BROWN SHOES THE BROWN SHOE CO. MAKERS
ST. LOUIS.
FOR DUPLICATE SEND 4¢ STAMPS TO THE BROWN SHOE CO. ST. LOUIS, MO.

Fig. 105. *Top to bottom, from left: Row 1,* Woman raising glass to lips, 1906, $750-1,000; Woman holding glass up, 1907, $750-1,000; *Row 2,* Woman facing viewer, people in background, 1920, $950-1,200; Woman in broad rim hat, facing left, 1910, $650-900; *Row 3,* Woman in broad rim had facing right, 1911, $600-900; Woman holding Coca Cola in left hand, pointing down, 1916, $800-1,000. *All courtesy of Roger Steckler.*

Drink Coca Cola

Fig. 106. From Coca Cola annual series, 1914, $600-700. *Collection of Dr. Robert Kalish, courtesy of Ted Hake's Americana.*

Fig. 107. *Top to bottom, from left: Row 1,* "Drink Coca-Cola" girl at table, 1904, exposition in background, "J. B. Carroll, Chicago," "Duplicate mirrors 5¢ postage, Coca-Cola Company, Atlanta, Ga." on edge, $500-700; *Row 2,* "Drink Coca-Cola" reg. U. S. Pat. Off (girl in yellow with yellow hat), key-chain attached, "1973 memorabilia, The Coca-Cola Company" on edge, $5-20; "Drink Coca-Cola" reg. U. S. Pat. Off. (girl in red with red hat), "1973 memorabilia, The Coca-Cola Company" on edge, $5-20; *Row 3,* "5¢, Drink Coca-Cola," "07003, Copyright 1904, by Detroit Photographic Co." on face, reproduction, $30-40.

Fig. 108. *Top to bottom:* French inscription "Deja des lunettes!" "Sa maman n'emploie pas des lampes Philips," translation: "Glasses already!" "Her mother doesn't use Philips bulbs," $100-150; Spanish inscription "Pildoritas de Reuter," "la mensajera de la salud," "impide autointoxicacion," entire translation: "Reuters little pills; for constipation, migraine, lack of appetite, etc.," "the messenger of health," "prevents self-poisoning." Spanish inscription on edge "Las Pildoritas de Reuter legitimas llevan en la etiqueta, La firma de Barclay & Co., copyrighted 1916 by Barclay & Co." (see Fig. 95), translation: "authentic Reuters Pills carry on the label, the firm of Barclay & Co.," Cuban, $100-150.

Fig. 109. *Clockwise from upper left:* Nature's Remedy, tablets, "Parisian Nov. Co., Chicago" on edge, $60-70; Cascarets, Sterling Remedy Co., Chicago or New York (boy on potty), $60-80; Cascarets, candy cathartic, Sterling Remedy Co., Wheeling, West Virginia, "Bastian Bros. Co., Rochester, N.Y." on edge (woman reclining on crescent moon "C"), $60-80; Angelus Marshmallows, "Mirror free with package of Angelus Marshmallows. Ask your dealer, or send 3-2¢ Stamps to Rueckheim Bros. & Eckstein, Chicago" on edge (angel with horn standing on box), $100-125; Ammon & Person, Baby Brand Butterine, "The Whitehead & Hoag Co., Newark, N.J." on edge (baby buttering toast), $300-500; Angelus Marshmallows, "Mirror free with package of Angelus Marshmallows. Ask your dealer, or send 3-2 cent stamps to Rueckheim Bros. & Eckstein, Chicago" on edge, $150-200.

Dutch Kids

Fig. 110. Van Camp's Pork and Beans, "The Whitehead & Hoag Co., Newark, N.J." on edge, $125-200.

Boys with Paper & Flour

Fig. 111. *Top to bottom:* The Boston Herald, The Sunday Herald, "Bastian Bros., Rochester, N.Y." on edge, $125-145; Ceresota Flour, Prize Bread Flour of the World, $60-85; Hammond Dairy Feed, Greatest Milk Producer on Earth, "Reading Ribbon Badge Co., Rdg., PA.," $45-75; Alliance Milling Co., Denton, Texas, Peace Maker, "The Flower of Flours," $45-65.

Fig. 112. *Clockwise from top:* Oliver Chilled Plow Works, South Bend, Indiana, U.S.A., "Bastian Bros. Co., Rochester, N.Y." on edge, $100-125; Walter E. Schmidt, member, Board of Assessors (no location), $40-50; Brockton Fair 1903 (no state), $60-75; Beeman's, the Original Pepsin Gum, $200-300; E. J. Roberts, 17 Exchange St. (no city, state), "Cruver Mfg. Co., Chicago" on edge, $35-45.

More Purmell Favorites

Fig. 113. *Top to bottom, from left: Row 1,* Beeman's Pepsin Gum, $300-400; Uncle Sam, Casper National Bank, Casper, Wyoming, "American Art Works, Coshocton, Ohio" on edge, $700; *Row 2,* Bromo Seltzer, "Minks Badge & Novelty Co., Balt, MD" on edge, $675; *Row 3,* Artesia Cream, "Whitehead & Hoag, Newark, NJ" on edge, $850; *Row 4,* Majestic Range, St. Louis, Missouri, $225-250; Sucrene Feed, "American Milling Company, Phila, PA" on edge, $550. *All courtesy of Burt Purmell.*

Fig. 114. *Top to bottom, from left: Row 1*, Red Rock Ginger Ale, "Duplicate of this mirror 6 cts in postage," The Red Rock Co., Atlanta, Ga., $800-950; Betsy Ross Shoes, Whitehead & Hoag, $750-800; *Row 2*, Superior Drill, "Duplicate for 3-2 cent stamps," Whitehead & Hoag, $750-800; Buckeye Fence, Bastian Bros., $375-450; *Row 3*, Philadelphia Lawn Mower, Whitehead & Hoag, $400-475; Apperson Auto, Cruver Mfg. Co., Chicago (see Figs. 7, 157), $650-700. *All courtesy of Burt Purmell.*

Fig. 115. *Clockwise from top:* Mahoney's Saloon, Philadelphia (Pennsylvania), tinted photograph of interior, "Cruver Mfg. Co., Chicago" on edge, $125-200; Duffy's Pure Malt Whiskey, "J. B. Carroll, Chicago" on edge, $80-125; Same as previous mirror but in round format, with "The Whitehead & Hoag Co., Newark, N.J." on edge, $50-75; Bailey's Pure Rye, Philadelphia, Pennsylvania, "The Whitehead & Hoag Co., Newark, N.J." on edge, $300-500; Adam Scheidt Brewing Co., Norristown, Pennsylvania, Valley Forge Special, Beer, $100-150.

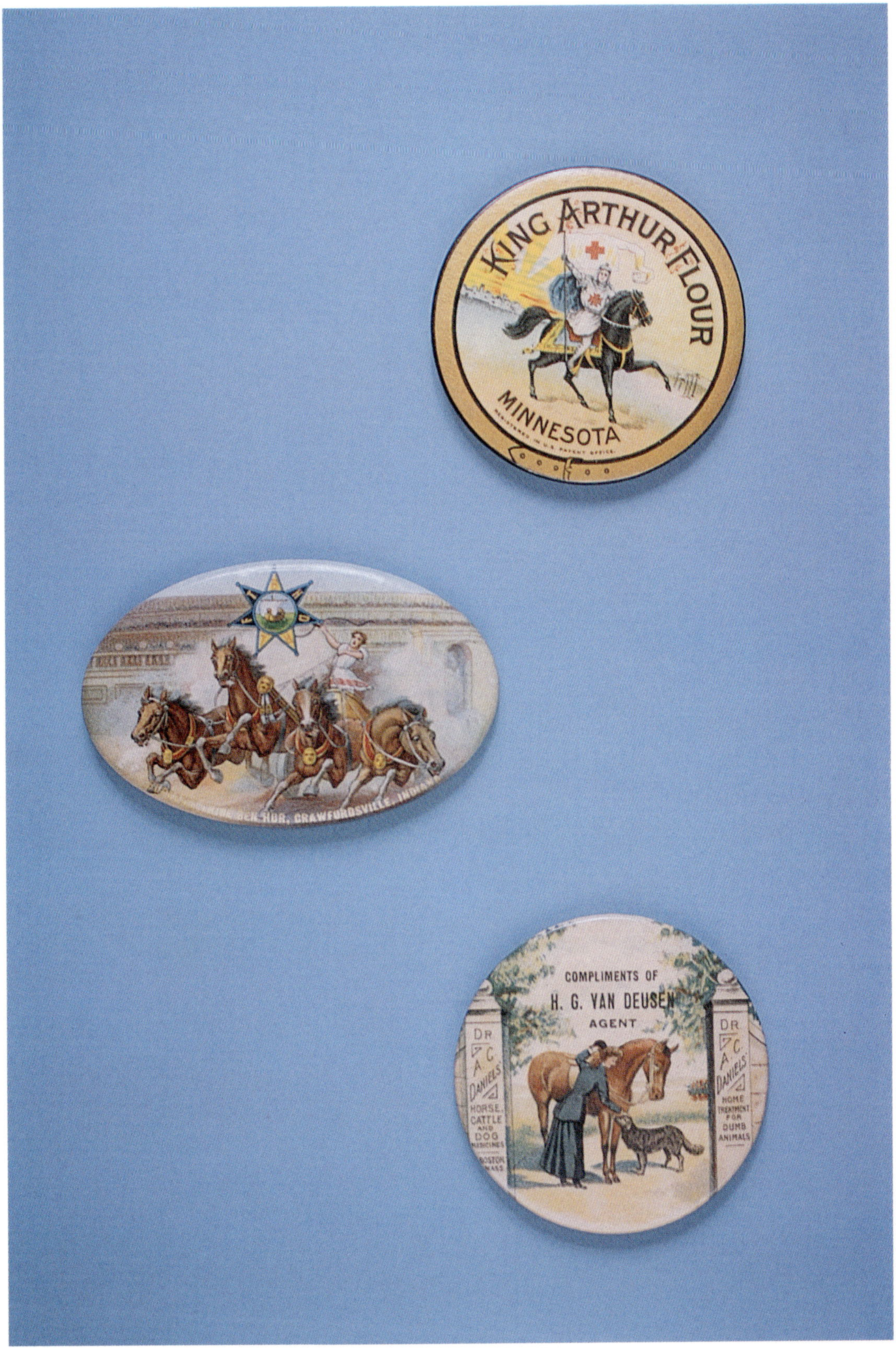

Fig. 116. *Top to bottom:* King Arthur Flour, Minnesota, "Bastian Bros., Rochester, N.Y." on edge, $100-125; Supreme Tribe Ben Hur, Crawfordsville, Indiana (FTHD, BHL in star), "The Whitehead & Hoag Co., Newark, N.J." on edge, $75-95; Dr. A. C. Daniels: Horse, Cattle and Dog Medicines, Home Treatment for Dumb Animals, compliments of H. G. Van Deusen, Agent, "Duplicate Mirrors 5¢ postage, Dr. A. C. Daniels, Boston" on edge, $100-125.

Fig. 117. *Top to bottom:* Wheeler Publishing Company, Chicago (Illinois), "Parisian Novelty Co., Chicago" on edge, $70-100, Victor "His Master's Voice," "The Whitehead & Hoag Co., Newark, N.J." on edge (see Fig. 123), $250-300; Mascot Crushed Cut Tobacco, "The Whitehead & Hoag Co., Newark, N.J." on edge, $50-75.

Fig. 118. *Top to bottom:* The Western Ohio Creamery Co., Greenville, Ohio, $40-60; Buffalo, New York, $35-50; Horlick's Malted Milk, "The Whitehead & Hoag Co., Newark, N.J." on edge, $50-80.

Fig. 119. *Top to bottom:* Peacock Coffee, \$80-125; Sample mirror, stock design, Hello Bill No. 52, with antlered deer, \$100-125; Old Reliable Coffee, Dutchman on dock, \$65-90. *All from the collection of Dr. Robert Kalish, courtesy of Ted Hake's Americana.*

Fig. 120. Liberty Mills, Vanity Self-rising Flour, $250-300. *Courtesy of Burt Purmell.*

Fig. 121. *Clockwise, from upper right:* Campbell's Tomato Soup, Joseph Campbell Preserve Co., Camden, New Jersey, U.S.A., 6 plates 10 cents, "Balto Badge & Nov. Co., Balto, MD." on edge, $175-225; Bell's Mocha & Java Coffee, J. H. Bell & Co., Chicago (Illinois), "Parisian Nov. Co., Chicago" on edge, $75-90; Red Seal Lye, Manufact'd by P. C. Tomason(?) & Co., Philadelphia, Pennsylvania, Disinfectant, Soap-Maker, Cleanser, "F. F. Pulver Co., Rochester, N.Y." on edge, $75-100; "Baked-Beans-Baked" with pork and tomato sauce, Lutz & Schramm Co., Pittsburgh, Pennsylvania, U.S.A., "Bastian Bros. Co., Rochester, N.Y." on edge, $40-65; Fort Bedford P-Nut Butter, "The American Art Works, Coshocton, O." on edge, $100-125; Woolsey's Best Copper Yacht Paint, Woolsey Paint & Color Co., Jersey City, New Jersey, U.S.A., "Bastian Bros. Co., Rochester, N.Y." on edge, $150-250.

Fig. 122. *Top to bottom, from left, Row 1,* Checkers Popcorn Confection, Shotwell Mfg. Co., Chicago (Illinois), "J.B. Carroll, Chicago" on edge, $100-125; White House Coffee, Divinell-Wright Company, Boston-Chicago, "Parisian Nov. Co., Chicago" on edge, $50-60; *Row 2,* Ex-Lax, A chocolate laxative, Ex-Lax Mfg. Co., New York, "The Whitehead & Hoag Co., Newark, N.J." on edge, $50-75; Golden Eagle Cigars, "The Whitehead & Hoag Co., Newark, N.J." on edge, $150-175; *Row 3,* Wisconsin Condensed Milk Co., New York, Lion Brand Condensed Milk, "Parisian Nov. Co., Chicago" on edge, $80-95; Lava Soap, Wm. Waltke & Co., St. Louis, U.S.A., "The Whitehead & Hoag Co., Newark, N.J." on edge, $50-60.

Fig. 123. *Top to bottom, from left: Row 1,* Welcome to Our Guest, State Council of Conn., Jr. O.U.A.M., Bridgeport (Connecticut), Sept. 27, 1910, $50-60; *Row 2,* Bjur Bros. New York, Pianos and Player Pianos, est'd 1887, A. W. Drew, Auburn, Maine, "The Whitehead & Hoag Co., Newark, N.J." on edge, $35-50; Symphony Hall Pops, G. E. Judd, Manager, Arthur Fiedler (pictured), conductor, Golden Jubilee – 1885-1935, $40-60; *Row 3,* Victor, "His Master's Voice," "The Whitehead & Hoag Co., Newark, N.J." on edge (see Fig. 117), $250-300; *Row 4,* Toscanini (portrait), "© Art Fair, 1968, N.Y. 10003" on edge, $10-20.

Fig. 124. *Top to bottom, from left: Row 1,* The Matchless Cunningham, John H. Heaton Piano Co., Camden, New Jersey, "Bastian Bros., Rochester, N.Y." on edge, $100-125; *Row 2,* Braunstein-Blatt Co., Atlantic City (New Jersey), Victor Red Seal record, "Celeste Aida," "Victrolas-Club Plan," "Bastian Bros. Co., Rochester, N.Y." on edge, $50-75; *Row 3,* Elmer's Kodaks-Victrolas, Watertown, New York, Red Seal Victrola record, "Her Bright Smile Haunts Me Still" (Edward Johnson), "©Reincke-Ellis Co., Chicago" on edge, $50-75; Geller & Seldin, Rockville Centre, Long Island (New York), "Her Bright Smile Haunts Me Still" (Edward Johnson), "©Reincke-Ellis Co., Chicago" on edge, $50-75; *Row 4,* Litholin waterproofed linen (collars), 25¢, McRae Haberdasher, Brattleboro, Vermont, "Parisian Novelty Co., Chicago" on edge, $50-80.

Fig. 125. *Top to bottom:* Football: Order your whiskey from The T. H. Haynes Co., Middlesboro, Kentucky, $45-75; Barrel: Belle of O. F. C., James A. Kelley, Rutland, Vermont (see Fig. 28), $65-85; Basketball: Horace Partridge Co., Boston-Cleveland-Charlotte (paperweight type), "The Whitehead & Hoag Co., Newark, N.J." on edge, $40-60; Cookie: all-celluloid, Huntley & Palmers, Osborne, design impressed into surface, $30-50.

Fig. 126. *Top to bottom,* Strunk & Moyer, Pigeon and Poultry Supply House, Reading, Pennsylvania, "Bastian Bros. Co., Rochester, N.Y." on edge, $75-95; (image of chick breaking out of eggshell), "Ehrman Mfg. Co., Chicago" on edge, $40-65; Peerless Poultry Supplies, Henry Albers Co., Los Angeles, $100-150; Ulmer's Special Breakfast Bacon, Jacob Ulmer Packing Co., Pottsville, Pennsylvania, "Bastian Bros. Co., Rochester, N.Y." on edge, $50-70.

Fig. 127. *Top to bottom:* Roelofs Hats, "The Whitehead & Hoag Co., Newark, N.J." on edge, $80-125; Chicago Tailoring Company (Chicago, Illinois), $80-100; Smoke Union Made Cigars (Sept. 1880, date on label), "The Whitehead & Hoag Co., Newark, N.J., American L.I.P. & B.A." on edge, $100-125.

Fig. 128. *Top to bottom, from left: Row 1,* Cathedral School from Park, Orlando, Fla. in center, $20-40; *Row 2,* Butterfly Sweet Shop, Gary, Indiana, $20-30; Bee Automobile Co., Allentown Pennsylvania, "G. Felsenthal & Sons, Chicago" on edge, $35-50; *Row 3,* Skeezix Shoes image in center, "Mfd. By sole licencees" and "Parisian Novelty Co., Chicago, Made in U.S.A." on edge, $75-95; *Row 4,* Geo. A. Perkins, Clothier, Mystic, Connecticut, "Patent applied for Cruver Mfg. Co., Chicago" on edge, $80-125; Jas. H. Tattersall, watchmaker and jeweler, Exeter, New Hampshire, "Patent applied for Cruver Mfg. Co., Chicago" on edge, $60-80; *Row 5,* W. M. Thayer, Jeweler, Belfast, Maine, "Copyrighted 1907, by Cruver Mfg. Co., Chicago" on edge, $100-150. All have borders of birthstones and their traits.

Jewelry Birthstone Borders

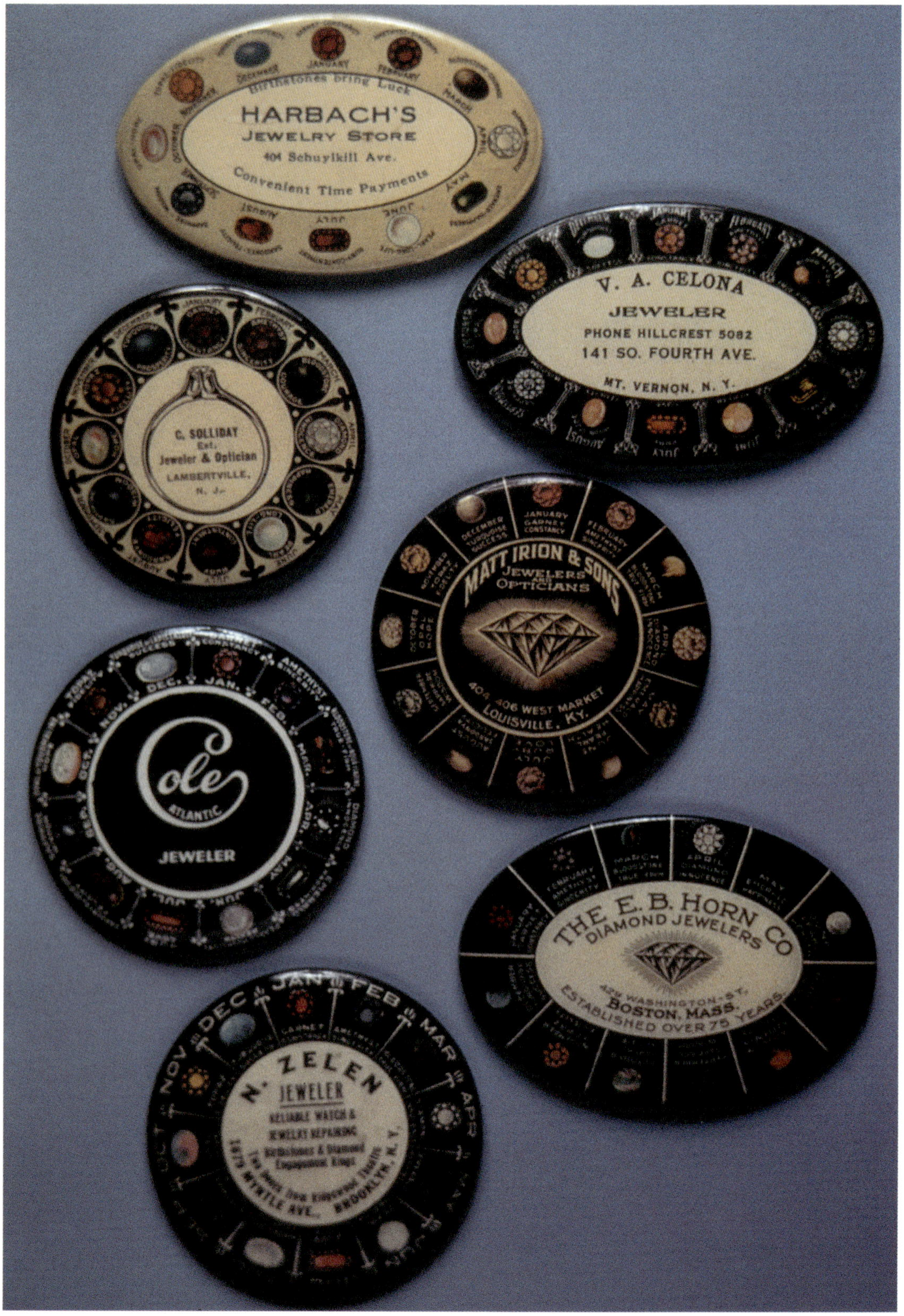

Fig. 129. *Clockwise, from top:* Harbach's Jewelry Store, Schuylkill Ave. (Philadelphia, Pennsylvania?), "Bastian Bros. Co., Rochester, N.Y." on edge, $25-35; V. A. Celona Jeweler, Mt. Vernon, New York, "Louis Bernheimer Co., N.Y." on edge, $25-35; Matt Irion & Sons, Jewelers and Opticians, Louisville, Kentucky, "Patented by Cruver Mfg. Co., Chicago, H.P.P." on edge, $50-60; The E. B. Horn Co., Diamond Jewelers, Boston, Massachusetts, "Copyrighted: 1907, by Cruver Mfg. Co., Chicago" on edge, $40-50; N. Zelen Jeweler, Brooklyn, New York, $30-40; Cole, Atlantic, Jeweler, "Parisian Novelty Co., Chicago, Made in U.S.A." on edge, $20-30; C. Solliday, Est. Jeweler & Optician, Lambertville, New Jersey, $40-60. All have borders of birthstones and their traits.

Fig. 130. *Top to bottom, from left: Row 1,* Alfred C. J. Williams, Furrier, Hartford, Connecticut, "Bastian Bros, Rochester, N.Y." on face, $25-35; Blank center, stock mirror, border of birthstones and traits, "The Whitehead & Hoag Co., Newark, N.J." on edge, $25-35; *Row 2,* Birthday greetings from Curtis Publishing Company, Philadelphia (Pennsylvania), The Saturday Evening Post, Ladies' Home Journal, Country Gentleman, "Copyrighted 1907 by Cruver Mfg. Co., Chicago" on edge, $30-45; The Pastime, Tilton, New Hampshire, $30-40; *Row 3,* A. B. Smith & Co., Factory, Boston (Massachusetts), Roman-faced clock border, "Cruver Mfg. Co., Chicago" on edge (see Fig. 6), $70-85; *Row 4,* The Fashion Specialty Cloak & Suit Store, Lewiston, Maine, "Patented by Cruver Mfg. Co., Chicago" on edge, $25-35; "Sla-Dale" Quality…Hosiery, Slatedale Knitting Mills Inc., Mills: Slatedale, Pennsylvania, Office: Harrisburg, Pennsylvania, $25-35. All but A. B. Smith & Co. have borders of birthstones and their traits.

Tape Measure Edging

Fig. 131. Continental Tailoring, Chicago, "Parisian Novelty Co., Chicago" on edge, $65-100.

More Birthstones & Traits

Fig. 132. *Top to bottom, from left: Row 1,* The Radcliffe Shoe for women, $25-30; Dunbar's Drug Store, Waterville, Maine, $25-30; *Row 2,* Allen & Van Dyke, Brooklyn, New York, large paperweight type, "The American Art Works, Coshocton, O." on edge (see Fig. 28), $25-50; *Row 3,* The Princeton Shoe Repairing Co., Princeton, New Jersey, $25-30; Coco-nut Margarine (Butterine), J. Rollick, Frankford (no state), $25-30. All have borders of birthstones and their traits.

Fig. 133. *Top to bottom, from left: Row 1,* Boot & Shoe Workers Union, Factory No.___, "Ehrman Mfg. Co., Boston" on edge, $30-50; *Row 2,* The Surprise, Ogdensburg, New York, Hood Rubbers, Hood's Kickoff Heel, with border of birthstones and traits, "Parisian Nov. Co., Chicago" on edge, $25-35; Slipknot Safety Heel, Plymouth Rubber Co., Canton, Massachusetts, "Made in America, Parisian Nov. Co., Chicago" on edge, $50-60; *Row 3,* J. D. LaCroix, South Hadley Falls, Massachusetts (same inner design as first mirror in Row 2, but with calendar border for 1912-1913), "Parisian Nov. Co., Chicago" on edge, $40-50; *Row 4,* F. W. Hahn & Co., wholesale distributors (no city, state), Shawmut Rubbers, "Bastian Bros. Co., Rochester, N.Y." on edge, $85-110; Emery, Bird, Thayer Dry Goods Co., Kansas City, Missouri, Shawmut Rubbers, with borders of birthstones and traits, $40-50.

Fig. 134. *Top to bottom:* John P. Deering for Governor (no state, no date), "The Whitehead & Hoag Co., Newark, N.J." on edge, $20-30; John F. Hylan for Mayor (no city or state) (NYC?), border of birthstones and their traits, "Made by Benj. Harris Co. on the Bowery 229, N.Y.C." on edge, $20-30; Worcester (MA?) needs Mary Scano, City Council, $5-10.

Fig. 135. Democratic Ticket, Vermillion County, and FDR, Daily and Sunday Times, $50-75. *Courtesy of Roger Steckler.*

Fig. 136. *Top to bottom, from left: Row 1,* Kirst & Son, Beautifiers of Homes, Reading (Pennsylvania), $30-50; The Ladies of the Maccabees, Headquarters World Maccabee Temple, Port Huron, Michigan, "Bastian Bros. Co., Rochester, N.Y." on edge (see Fig. 138), $50-75; *Row 2,* G. C. Fishbach, Music Dealer, Allentown, Pennsylvania, Oct. 19, 1899, unveiling of Soldiers and Sailors Monument, "The Whitehead & Hoag Co., Newark, N.J." on edge (see Fig. 144), $30-50**;** *Row 3,* Missouri-Denver, 1908, S. G. L. (railroad?), "The Whitehead & Hoag Co., Newark, N.J." on edge (reverse has same decorative framing as Fig. 162; also see Fig. 158**),** $35-45; Citizens Fire Co., No. 1, Tamaqua, Pennsylvania, 13th Annual Convention, June 19-22, 1916, image of fire engine and station in center, "Ehrman Mfg. Co., Boston" on edge (see Fig. 156), $35-50.

Bicycles

Fig. 137. National Bicycles, National Cycle Manufacturing Co., Bay City, Mich., Whitehead & Hoag, Newark, N.J. on edge (see Fig. 14), $425-475. *Collection of Mark and Lynda Suozzi, Ashfield, Mass.*

Fig. 138. *Top to bottom, from left: Row 1,* "Clarite," Columbia Tool Steel Co., "Greenduek, Chicago" on edge, \$40-50; *Row 2,* Patriot & Field Day, Monday, May 31st (no year), "The Whitehead & Hoag Co., Newark, N.J." on edge, \$50-75; *Row 3,* Woman's tinted portrait in patriotic stock framing, \$80-125; WWI soldier's tinted portrait in same stock framing with gilded oval separation, \$80-125; *Row 4,* The Allies, WWII nations, "Bastian Bros., Rochester, N.Y." on edge, \$40-50.

Historic/Military

Fig. 139. 18th Anniversary Camp No. 678, George Washington portrait with patriotic and military images, $100-150.

Fig. 140. *Top to bottom, from left: Row 1,* The Capitol, Washington, D.C., $20-30; *Row 2,* Statue of Liberty, New York Harbor, New York, "Bastian Bros. Co., Rochester, N.Y." on edge, $125-175; Liberty Bell, Philadelphia, Pennsylvania, "Bastian Bros. Co., Rochester, N.Y." on edge, $60-80; Niagara Falls, New York (see Fig. 54), $40-50; *Row 3,* The Capitol, Washington, D.C., $20-30.

A Patriotic Advertisement

Fig. 141. Dixie Tailoring Company, Chicago, Ill, Statue of Liberty, "E. W. Smock & Co., Chicago" on edge, $200-250.

Fig. 142. *Top to bottom:* Kaufmann's Navajo Tribe, WRAW 6:30 p.m., Chief Brown Wolf, Mon, Wed, Fri, $200-300; Souvenir of the Registering and Drawing for the Tripp County Lands, Rosebud Reservation, Dallas, South Dakota, October 5-19, 1908, "Mathis-Mets Co., Dubuque, IA." on edge, $300-400; Samoset Chocolates, trademark, $100-200.

Fig. 143. *Top to bottom, from left: Row 1,* The Oranges in Jersey, East Orange & Ampere Land Company, "J.B. Carroll, Chicago" on edge, $65-75; *Row 2,* W. W. (Apple) & Son, Jewelers & Opticians, Lancaster, Pennsylvania, "The Coshocton Specialty Co., Coshocton, Ohio" on edge, $70-90; Apple Butter, "L & S trademark," Lutz & Schramm Co., Pittsburgh, Pennsylvania, "Arthur von Senden Co., Pittsburgh" on edge, $50-60; *Row 4,* Los Angeles, County of Fruit and Flowers (with gold foil beneath, transparent overlay), "The Whitehead & Hoag Co., Newark, N.J." on edge, $80-125.

Memorials & Monuments

Fig. 144. *Top to bottom, from left: Row 1,* Soldiers' & Sailors' Monument, Washington Park, Rochester, New York, $20-30; Pilgrim Hall (no city or state), $20-30; *Row 2,* William McKinley Memorial, Canton, Ohio, "The Whitehead & Hoag Co., Newark, N.J." on edge, $35-75; Memorial Church, Stanford University, California, $20-30; *Row 3,* Library of Congress, Washington, D.C., $30-40; Pennsylvania State Memorial, Gettysburg, Pennsylvania, $30-40.

Fig. 145. *Clockwise from upper right:* Natatorium, Pool 100 by 30 Feet, Reading, Pennsylvania (reducing mirror), "Julius Weber, Reading, Pa" on edge, \$50-100; Old Stone Mill, Newport, Rhode Island, \$25-35; "High Rock, 2000 ft. Above Sea Level" (no city or state), "W.P., WL" (?) signed, early all-celluloid rim, \$20-30; Cargill Falls, Putnam, Connecticut, \$20-30; Souvenir 1905, Birds Eye View, Chestnut Hill Park, Philadelphia, Pennsylvania, "The Whitehead & Hoag Co., Newark, N.J." on edge, \$30-50.

Fig. 146. *Top to bottom, from left: Row 1,* Palace of Machinery, The World's Fair, St. Louis, 1904, in brass frame with handle, possible edge inscription inside frame, $100-125; *Row 2,* Canadian National Exhibition, Toronto, Aug. 27 to Sept. 10, 1921, "Bastian Bros. Ltd., Toronto" on edge, $25-35; Electric Tower, Pan-American Exposition, 1901, $30-40; *Row 3,* The Mighty Dollar Cigar Store & Café, Rochester, New York, "The Whitehead & Hoag Co., Newark, N.J." on edge, $25-35; *Row 4,* Geo. Ehrenfried Co., Fiftieth Anniversary (clothier, department store? No city or state), in brass frame with handle, possible edge inscription inside frame, $30-40; Welcome to Lewiston, Annual Convention Maine State Grange, Dec. 21-22-33, 1920, "Bastian Bros. Co., Rochester, N.Y." on edge, $25-35.

Fig. 147. *Top to bottom, from left: Row 1,* The Heald Chain of Schools, Fresno, Chico, San Francisco, Reno, Ocean Park, Santa Cruz, Riverside, Stockton, Oakland, San Jose, Long Beach, Los Angeles, "J.B. Carroll, Chicago" on edge, $30-50; Breinig & Bachman, clothiers, furnishers, Allentown, Pennsylvania, "The Whitehead & Hoag Co., Newark, N.J." on edge, $30-50; *Row 2,* C. K. Whitner & Co., Reading, Pennsylvania, $30-50; Berry Brothers, Ltd., Varnish M'fr's., Castle Copal, Pan-American Exposition, Buffalo (New York), 1901, "The Whitehead & Hoag Co., Newark, N.J." on edge, $45-60; *Row 3,* Hotel Elberon and Fireproof Annex, Atlantic City, New Jersey, Robert B. Ludy, M.D., owner and propr., "T. Gleason, Brooklyn, N.Y.(?)" on edge, $30-50; "Home of the Review, Alliance's Best Newspaper" (no state), $30-50; *Row 4,* New Home of Enterprise Custom Laundry, Lynn Massachusetts, $30-50; The Praetorians Scientific, Fraternal Insurance, Praetorian Building, Dallas, Texas, $30-50.

Making Progress

Fig. 148. *Top to bottom, from left: Row 1,* Hotel Statler, Buffalo, New York, $200; *Row 2,* Dowagiac Drills and Seeders, "Cruver Mfg. Co., Chicago, Ill" on edge, $800; Gibson Furniture Company, "Bastian Bros, Rochester, NY" on edge, $750; *Row 3,* Keith Clothing, $275. *All courtesy of Burt Purmell.*

Stores, Commercial Buildings, Homes

Fig. 149. *Top to bottom, from left: Row 1,* Norumbega Park Restaurant, Auburndale, Massachusetts, U.S.A., $25-40; The Old Reliable Store, Kline, Eppihimer & Co., Penn Square, Reading, Pennsylvania (with flag), $25-40; *Row 2,* General Offices, Knights of the Maccabees, Of the World, Detroit, Michigan, "Duplicates mailed for three 2¢ stamps, to general offices, Detroit, Mich." on edge, $80-100; *Row 3,* "It's up to you to get a home like this in North Dakota," $25-40; "Forest House, Youngsville, N.Y.," $25-40; *Row 4,* Pillsbury & Baldwin Co., St. Johnsbury, Vermont, "Cruver Mfg. Co., Chicago, U.S.A." on edge, $25-40; Weinsheimer Hardware Co., Walnut St. (Philadelphia, Pennsylvania), $30-60.

Fig. 150. *Top to bottom, from left: Row 1,* Old Home Week, Chelsea, Vermont, Marcellus Gould & Hilas E. Roberts, 1915, $15-30; *Row 2,* St. Joseph's Church, Amesbury, Massachusetts, Rev'd F. Lee, Pastor, "The Whitehead & Hoag Co., Newark, N.J." on edge, $15-30; Souvenir of Old Home Week, Mechanic Falls, Maine, "The Whitehead & Hoag Co., Newark, N.J." on edge, $15-30; Chelsea Old Home Week, August 1925, $15-30; *Row 3,* St. Peter's Ev. Luth. Church, Rittersville, Pennsylvania, Rev. I. B. Ritter, $15-30; The Orange Home, Hatboro, Pennsylvania, $15-30; *Row 4,* St. Luke's Union Church, Inc., Dedication June 14th, 1931, Shoemakersville, Pennsylvania, $15-25; Jacob's Union Church, Jacksonville, Pennsylvania, Rev. A. D. Ebert and Rev. C. R. Rahn, dedicated 1927, "Bastian Bros., Rochester, N.Y." on edge, $15-30.

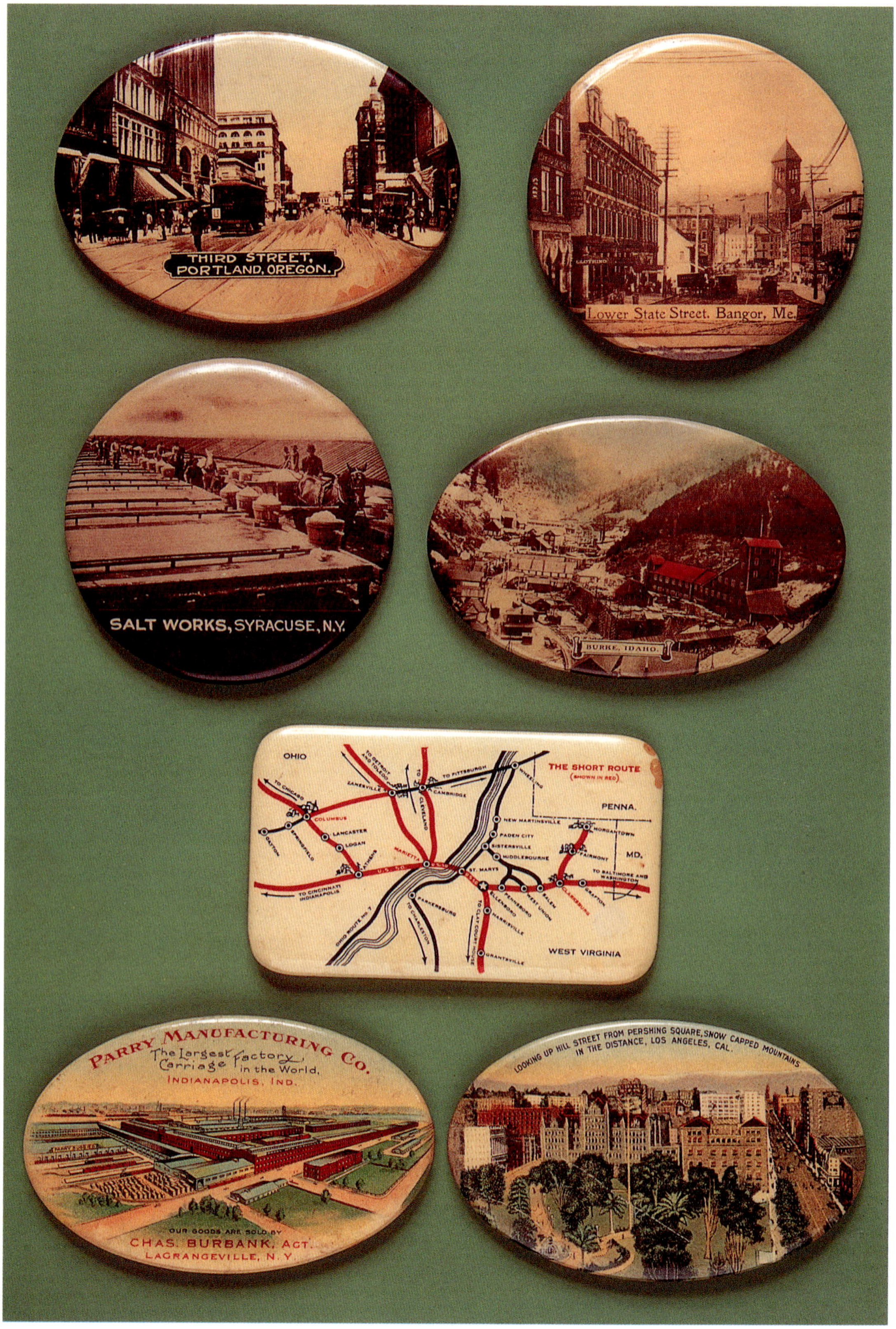

Fig. 151. *Top to bottom, from left Row 1,* Third Street, Portland, Oregon, $30-40; Lower State Street, Bangor, Maine, $30-40; *Row 2,* Salt Works, Syracuse, New York, $30-40; Burke, Idaho, $30-40; *Row 3,* Map of The Short Route, Ohio, West Virginia and Maryland, "Cruver Mfg. Co., Chicago" on edge, $100-150; *Row 4,* Parry Manufacturing Co., Indianapolis, Indiana, carriages sold by Chas. Burbank, Agt., LaGrangeville, New York, "Bastian Bros. Co., Rochester, N.Y." on edge, $90-150; Looking up Hill Street from Pershing Square, snow capped mountains in the distance, Los Angeles, California, $30-40.

Fig. 152. *Top to bottom,* A town with distant mountains (unidentified), $20-30; Eva's Farm, Purling, New York, "H.L. Moore Co., West Yarmouth, Mass" on edge, $20-30.

Fig. 153. *Top to bottom, from left, Row 1,* Scranton Stove Works, Scranton, Pennsylvania, "The Meek Co., Coshocton, O." on edge, $70-80; *Row 2,* Sunshine and Othello Enameled Ranges, The Reading Stove Works, Orr, Painter & Co., Reading, Pennsylvania, "Bastian Bros. Co., Rochester, N.Y." on edge, $30-40; *Row 3,* Garland Stoves and Ranges, "Parisian Novelty Co., Chicago" on edge, $30-40; *Row 4,* Hersey Furniture Co., Worthington St. (no city or state), Hub Ranges, "Cruver Mfg. Co., Chicago, U.S.A., ICP" on edge, $60-80; Seltzer & Sons, Somerville, N. J., Buckwalter Stove Co., Royersford, Pennsylvania, $75-90.

Fig. 154. *Top to bottom, from left: Row 1,* The Monarch Typewriter Company, Syracuse, New York, U.S.A., Sales Office, Albany, New York (reducing mirror on reverse), "The Whitehead & Hoag Co., Newark, N.J." on edge, $60-90; *Row 2,* Remington Standard Typewriter, "The Whitehead & Hoag Co., Newark, N.J." on edge, $100-150; The Smith Premier, Syracuse, New York, U.S.A., "The Whitehead & Hoag Co., Newark, N.J." on edge, $60-90; *Row 3,* Large, weighted paperweight type, same as first mirror in Row 2 except: "Parisian Nov. Co., Chicago" on edge, $40-60.

Fig. 155. *Top to bottom, from left, Row 1,* Starrett Tools, "Bastian Bros. Co., Rochester, N.Y." on edge, $50-60; The "New-Way" Air Cooled, Lansing, Michigan, U.S.A., "Parisian Novelty Co., Chicago" on edge, $75-100; *Row 2,* Sangamo Direct Current Meter, Sangamo Electric Co., Springfield, Illinois, "Parisian Nov. Co., Chicago, U.S.A." on edge, $80-100; *Row 3,* The O. & J. Machine Co., Worchester, Massachusetts, Automatic Labeler, Automatic Ten Head Crowner (paperweight type), "J.B. Carroll, Chicago" on edge, $50-75.

Fig. 156. *Top to bottom:* Bob White, Frosted Foods, Ardmore (Pennsylvania), $75-100; G. & J. Tires, Pittsburgh Rubber Co., Pittsburgh (Pennsylvania), $50-75; "Mack" leading gasoline truck in America, paperweight type, "Parisian Nov. Co., Chicago" on edge, $100-200.

Early Taxi

Fig. 157. Portland Taxicab Co., Maine (see Figs. 7, 114), $150-175. *Courtesy of Roger Steckler.*

Fig. 158. *Top to bottom, from left: Row 1,* The Travelers Insurance Company, Hartford, Connecticut, "The Railroad Men's Reliance," "The Whitehead & Hoag Co., Newark, N.J." on edge, $80-100; *Row 2,* Mt. Tom Railroad, Holyoke, Massachusetts, view near lower station, "The Whitehead & Hoag Co., Newark, N.J." on edge, $30-50; Parkesburg, Pennsylvania, Philadelphia Division, Veterans Association, Sept. 16, 1911, "The Whitehead & Hoag Co., Newark, N.J." on edge, $100-150; *Row 3,* Frisco System, $90-125; *Row 4,* Carter's Overalls, H. W. Carter & Sons, Lebanon, New Hampshire, "The Whitehead & Hoag Co., Newark, N.J." on edge, $60-80.

Fig. 159. Cunard White Star Liner, "Queen Elizabeth," $100-125. *Courtesy of Roger Steckler.*

Fig. 160. "American Line, Philadelphia, Liverpool, Queenstown," liner with three flags and name "Queen"(?) on face, $100-125.

Riverboats . . .

Fig. 161. *Top to bottom, from left: Row 1,* Rutland Business College, L. J. Egelston, Principal (Rutland, Vermont), "Brown & Bigelow, St. Paul (Minn)" on edge, $100-125; S. S. Jack O'Lantern of Boston (Massachusetts), $40-60; *Row 2,* "Hendrick Hudson," steamship with flags, $75-90; "Robert Fulton," steamship with flags, $75-90; *Row 3,* "Gov. Andrew," steamship, $40-60; *Row 4,* 13th Annual Convention of the New Hampshire State Fireman's Assn., Meredith, New Hampshire, Sept. 28, 1910, "The Whitehead & Hoag Co., Newark, N.J." on edge, $60-70; Moores Muresco (paint), Gould and Cutler, Boston (Massachusetts), "J.B. Carroll, Chicago" on edge, $40-60.

Fig. 162. Reverse of New Hampshire State Fireman's Assn., from Fig. 161 (also see Fig. 136).

. . . A Shipshape Beauty!

Fig. 163. "Going to Tolchester Beach," "The Whitehead & Hoag Co., Newark, N.J." on edge, $300-350.

Fig. 164. *Clockwise from top:* The Kraissl Company, Hackensack, New Jersey, pumps-strainers, engineering equipment, "Whitehead & Hoag, Newark, N.J." on edge, $30-40; Tower Milling Co., Greenville, Michigan, "Parisian Novelty Co., Chicago" on edge, $50-70; Gribben Lumber Company, "Coller McLean Co., Straw . . . (illegible)" on edge, $25-40; Bagley Old Colony Tobacco, John J. Bagley & Co., Detroit, Michigan, "Bastian Bros Co., Rochester, N.Y." on edge, $60-80.

Fig. 165. *Top to bottom:* Dr. Swett's original Root Beer, Boston, Massachusetts, "The American Art Works, Coshocton, O." on edge, $300-400; Ste. Marie's, Lewiston, Maine, built by L. Malo & Sons, gen. contractors, Lewiston, Maine, "Cruver Mfg. Co., Chicago" on edge, $20-30.

Fig. 166. W. Atlee Burpee & Co. (no location), shown on lime and maroon backgrounds, $75-95.

Fig. 167. *From left*: Jerry, Grayson, and Cynthia Dantzic, pinback button, made on same machine that produces pocket mirrors, at New England antiques fair, c. 1969.

Fig. 168. Federal Express, ZAP Mail, 2000, \$5-10. *Courtesy of Christina and Irwin Rubin.*

Endnote

Never the End . . .

Although this had seemed a good stopping place, how could I not include these just-acquired treasures:

- a full compact, top and bottom, (definitively identifying the Lydia Pinkham and others as container tops)
- a coin inclusion mirror, containing a near mint 1930 penny!
- a rectangular Campbell's Soup
- another fine barrel shaped mirror . . .

. . . and the final mirror, a recent, but nevertheless historic memento (just presented to me by a student in my Color class at Cooper Union), with the Twin Towers still holding their place in the New York skyline—a fine example of the difference between value and desirability!

Fig. 169. *Top to bottom:* Ritz Credit Jewelers, Louisville, Kentucky, top of compact (see Figs. 20, 27), $75-100; bottom of compact (see above); Re-elect William V. Pacelli, Alderman, coin inclusion mirror, "Allied Printing Trades Council, Union Label" on edge, with 1930 penny, $75-125.

So Long, For Now . . .

Fig. 170. *Top to bottom:* Campbell's Soups, Joseph Campbell Company, Camden, New Jersey, U.S.A. (see Fig. 121), $75-125; Home Oil Co., oil, gasoline and lubricants, St. Paul, Minnesota (see Fig. 125), $75-125; New York, NYF-07, Distributed by Widthman (USA) Inc., made in the U.S.A., Brooklyn Bridge with Twin Towers in skyline, $5-10.

Glossary

In addition to several basic terms, I have included a number of phrases used by dealers and collectors to help you become more conversant with the shop talk, or idiom, of the dedicated pocket mirror maven.

BenDay process. Used in printing to achieve tonal variation by placing a graded dot screen over art work before making the printing plate.

Celluloid. Material made of cellulose nitrate and camphor, the first "plastic," used for most mirrors after the turn of the twentieth century.

Cherry-picking. Scanning a mixed-price group of mirrors and selecting the best value.

Collet. The metal loop or ring, a clamp that holds all parts of the mirror in place.

Copy. The actual letters or wording on a design.

Curl. The celluloid rim or edge of the mirror on which the maker's name is often printed— extends underneath, up to the collet.

Face. The picture side, or front, of the mirror.

Fake. An imitation, reproduction, passed off as an original work.

Fantasy. Invented or created to appear authentic, but not so in fact.

Field. The area around the main or central image, often containing type or lettering.

Fish. A buyer unfamiliar with the market value, willing to pay more than the item is worth.

Foxing. Rust stain resulting from moisture at the collet's edge, or through seepage beneath a crack or tear in the celluloid.

Good for. A mirror with an inscription stating that it is good for (may be exchanged or shown for) a certain amount of cash or a free product.

Hole-filler. A less than acceptable specimen of a desired mirror, filling a space in a collection until it can be upgraded.

Maverick. A mirror identifying a product or company, but with no city or state named.

Mint. Perfect, "day it was made" condition.

No fixed value. Balancing the price asked against what you are willing to pay, considering the variables of the moment.

Novodel. A fantasy used to deceive the buyer.

Portrait. The main design on the mirror's face, not necessarily a person's face; often a product or other pictorial image.

Repro. A reproduction, fake or imitation. This is to be avoided.

Shark. Knows a piece is overpriced, and negotiates . . . carefully.

Shell. An early embossed metal mirror facing, usually brass or silver-plate, sometimes copper, aluminum, or nickel.

Stock. A standardized image on a mirror blank, to be overprinted with a specific company name or merchant's copy.

Upgrade. To trade a "hole-filler" for a more desirable mirror.

Bibliography

Books (all include material on pocket mirrors)

Fertig, Howard, and Flo Fertig. Advertising Giveaways. In *The Encyclopedia of Collections*. Alexandria, Virginia: Time-Life Books, 1978.

Hake, Ted, and Russ King. *Price Guide to Collectible Pin-Back Buttons 1896-1986*. Iola, Wisconsin: Krause Publications, 1986, 1991.

Kaduck, John M. *Collecting Advertising Mirrors*. Cleveland, Ohio: John M. Kaduck, 1973.

Kovel, Ralph, and Terry Kovel. *Kovels' Antiques & Collectibles Price List*. New York: Crown Publishers, Inc. (annually).

Liman, Ellen, with Lewis Liman. Advertiques, Reflecting the Past: Mirrors. In *The Collecting Book*. N.p.: Penguin Books, 1980.

Petretti, Allan. *Petretti's Coca Cola Collectibles Price Guide (10th edition)*. Iola, Wisconsin: Krause Publications, 1997.

Catalogues

Dunn, Hal V. *Catalog of Western Good For Trade Mirrors*. Granada Hills, California: Koppenhaver Press, 1995.

Hake's Americana & Collectibles (five annual auction catalogues, since 1967). Box 1444, York, Pennsylvania, 17405-1444.

Moffat, Bob (six mail-bid sale catalogues per year). P.O. Box 281, Auburn, Massachusetts, 01501.

The Personal Collection of Roger Steckler: Pocket Mirrors, Tip Trays. Timonium, Maryland: Richard Opfer Auctioneering, Inc., November 24, 1997.

Rex Stark Americana (approximately three catalogues per year), P.O. Box 1029, Gardner, Massachusetts, 01440.

Periodicals

American Heritage, 24, no. 1 (December, 1972).

Antique Trader, Iola, Wisconsin (weekly newspaper format publication listing shows, auctions, dealers, shops, calendar of events, classifieds, and other articles relating to antiques and collectibles, nationally).

Gaylord, Bill. "Advertising Pocket Mirrors: Small Treasures." *Tin Type*. 20, no. 5 (August, 1991): 3-17.

Purmell, Burt. "Collecting Now – Pocket Mirrors." *Country Living*, June 1992, 62.

"Collecting Advertising Pocket Mirrors." *Inside Collector*, May 1974, 81.

Spiller, Dr. Burton. "Reflections of our Past, Advertising, Pocket Mirrors." *Spinning Wheel*, January-February, 1974.

Treasure Chest, Providence, Rhode Island (monthly newspaper format publication listing shows, auctions, dealers, shops, calendar of events, classifieds, and other articles relating to collectibles in the Northeast, from Maine to Maryland, approximately 36 pages per issue).

Index